Manfred Velden

Human-like Computers

A Lesson in Absurdity

Schwabe Verlag

MIX
Papier aus verantwor-
tungsvollen Quellen
FSC® C083411

Bibliographic information published by the Deutsche Nationalbibliothek
The Deutsche Nationalbibliothek lists this publication in the Deutsche Nationalbibliografie;
detailed bibliographic data are available on the Internet at http://dnb.dnb.de.

© 2022 Schwabe Verlag, Schwabe Verlagsgruppe AG, Basel, Schweiz
This work is protected by copyright. No part of it may be reproduced, stored in a retrieval system or transmitted in any form or by any means, electronic, mechanical, photocopying, recording, or otherwise, or translated, without the prior written permission of the publisher.
Cover design: icona basel gmbH, Basel
Cover: Kathrin Strohschnieder, STROH Design, Oldenburg
Graphic design: icona basel gmbh, Basel
Typesetting: Schwabe Verlag, Berlin
Print: CPI books GmbH, Leck
Printed in Germany
ISBN Print 978-3-7965-4525-2
ISBN eBook (PDF) 978-3-7965-4526-9
DOI 10.24894/978-3-7965-4526-9
The ebook has identical page numbers to the print edition (first printing) and supports full-text search.
Furthermore, the table of contents is linked to the headings.

rights@schwabe.ch
www.schwabe.ch

Contents

Preface 7

Introduction 9

Whole fields of science gone awry 13
    Planetary movements – a theological perspective 13
    The heritability of intelligence –
    a scientific zombie still going strong 14
    Behaviorism – a mindless psychology 16
    Psychsomatics – an etiology lost and regained 18
    Mirror neurons – a debacle caused by hype 20
    Localization of mental traits – phrenology's long goodbye 21

AI peculiarities 27
    AI – a misnomer 27
    Misleading use of analogies – raison d'être of the field 29
    Predictions masquerading as facts 30
    A ghost in everything 32

Human-like computers – a belief system 35

**The phenomenology of mental functions –
a neglected issue in psychology** 37

The biological basis of mental functions 43

Does "behave like" mean "*be* like"? 53

The brain is not a computer 55

The Turing Test – what can it tell us? 65

Taking the brain for the human – a deprived view 71

A short phenomenology of mental functions 73
    Memory 74
    Perception 81
    Motivation and emotion 85
    Learning 90
    Consciousness 92

Human-like computers – a scientistic delusion 107

Acknowledgements 119

Literature 121

Index 125

Preface

When in the debate about the possible humanness of computers the MIT computer scientist Joseph Weizenbaum was hard pressed to concede that computers can be "socialized", he found that they could but in an extremely limited sense. He then stated:

> If both machines and humans are socializable, then we must ask in what way the socialization of the human must necessarily be different from that of the machine. The answer is, of course, so obvious that it makes the very asking of the question appear ludicrous, if indeed not obscene. It is a sign of the madness of our time that this issue is to be addressed at all. (Weizenbaum, 1976, p. 210)

For Weizenbaum the answer was so obvious that he even did not bother to give it. He probably thought that, as science progressed, the answer would eventually become obvious to everyone in the field anyway. But this did not happen. The opposite happened. Today, nearly half a century later, it is common "knowledge" among scientists and in a wide public that computers will, sooner or later, be human-like. Some think that computers already are.

So was Weizenbaum eventually proven wrong? Was he, who reasoned about the future of the computer world, about the "digital age", as thoroughly as nobody else, unable to foresee a future where the difference between humans and machines would become blurred, as many in the AI community want to make us believe, a development which, if it actually happened, would have a transforming impact on the future of mankind? In the decades following his above statement, Weizenbaum actually never distanced himself from it. Quite to the contrary, he was more and more convinced that a serious misdevelopment was occuring in man's attitude toward computers. So before we discuss his decades old statement as one caused by lack of fantasy and made obsolete by scientific progress, we had better take pains to delve into the scientific-philosophical depths of the man-machine issue. This is what this book is about. Unfortunately, this issue is not only, and predictably so, an extremely complex one, but, more importantly, the whole debate is also plagued by a deep ignorance about the nature of human mental processes, about the human psyche. This need not surprise us, because the debate is largely dominated by engineers, mathematicians and information scientists. Sadly, psychologists, actually the specialists responsible for the very mental functions the computer scientists intend to emulate, have contributed little to instructing those computer scientists about the nature of human mental functions or the grave definitional problems that plague the science of psychology. They rather decided to ride the computers-can-be-human-like wave and thus furthered rather than criticized that flawed project, promulgating even such outlandish ideas as that of computers becoming superintelligent or even superhuman, and that they eventually will take over control of the world from those outmoded creatures that used to be called humans.

Introduction

When planning this book, in which I want to refute the idea that there can be processes in machines comparable to mental processes in humans, i.e. that machines can be human-like, I first thought to take typical examples of computer functions allegedly comparable to human mental functions, compare the two and then decide whether it makes sense to regard computer functions as really human-like. But I soon realized that my task would be comparable to Sisyphus'. With the latest computer accomplishment popping up on a monthly if not weekly basis, I would have to start pushing the rock uphill anew every month or week with no end in sight. The book would have to be rewritten at least once a year. Machines with human-like mental functions now have been predicted for many decades (since the advent of electronic computers) and I am afraid that this will go on for many decades more unless the scientific community realizes that there are qualitative differences between human mental functions and functions running on computers, which cause quantitative comparisons (as with, for example, intelligence) to make no sense whatsoever. So the book will be about *a priori grounds* on which computers cannot be human-like.

When, long ago, I read Weizenbaum's comment that to compare the socialization of a machine with that of a human is a sign of madness, I immediately felt that this was the ultimate comment possible and did not expect Weizenbaum to explain why (which he did not bother to do anyway). But the insight into this madness not having befallen the scientific community nearly half a century later that "Why" must finally be delivered.

A comparison between the socialization of a human and that of a computer is not seen as mad in the AI community. This is so not because obvious similarities could be shown to exist between the two (there aren't any; there is just a faint analogy in that both are somehow affected by their environment) but simply because this kind of speak (and think) has become quite common among AI researchers. Take a typical phrase like "culture is poured into artificial brains" (Collins, 2018, p. 173). You can pour culture into computers as little as you can pour the theory of relativity into a teacup. Any nonsense can be brought forward in a gramatically correct form and, if done routinely, it may not be felt as nonsense any more. The above phrase about how culture can manage to get into a machine reminds one in its nonsensicality of Francis Picabia's[1] "The head is round in order to allow the thoughts to change direction". At a closer look much of AI speak appears to be inspired by Dadaism. The only difference between AI and Dada speak is that Dada is funny and the Dadaists consciously invented nonsensicalities in order to provoke and entertain, while AI speak is by no means funny and the AI researchers are convinced that they talk reasonably. So when reading the AI literature, we are confronted with all kinds of absurdities, like machines being socialized, having human feelings, motives, consciousness and religion, westernized human-like desires, and culture being poured or fed into machines, or with artificial slaves with human-like bodies, or with self replicating intelligent machines. I think that large parts of the AI literature may be seen as lessons in absurdity, presented as rational, scientifically based predictions about our digital future.

The idea of human-like computers is based, quite like its 18th century predecessor, *L'homme machine* (Man a Machine) by de la Mettrie (1921), on a simple mechanistic-materialistic belief, abandoned in physics long ago. As a consequence of this simplistic man-a-machine view, the whole debate about human mental functions in computers suffers, as mentioned, from a deep ignorance about the nature of human mental processes, about the human psyche. In addition to that, the engineering perspective of the field largely underestimates our deep ignorance about the biological, largely neural, basis (the biological substrate) of those processes. So I will in some detail deal with that substrate and then try to paint a realistic picture of the daunting complexity of human mental functions in accord with the simple motto "if you want to mimic something you should know what it is."

That picture will constitute the largest part of the book. My view of the man-a-machine matter is not just that the idea is a somehow problematic one, but that it is an absurd, i.e. a ridiculousy nonsensical one. This view,

---

1   Francis Picabia (1879–1953), painter, Dadaist, surrealist.

identical to the one Weizenbaum took half a century ago, today is one held by a small minority facing a huge majority of believers in human-like computers, both in a wide public and also among scientists. To convince the reader to join a small minority against a seemingly overwhelming opposition is no easy task. As a kind of psychological support in this task let me, before coming to the actual topic of the book (the absurdity of the man-a-machine notion), present some examples from mental history where a large majority had it wrong for decades, centuries or even milennia. And let me also point to some peculiar characteristics of the debate which must make us doubt that we are dealing with a scientific one in the first place. In terms of argumentation, the field must be seen as one following entirely its own rules.

Whole fields of science gone awry

Planetary movements – a theological perspective

When a Native American finally returned to his tribe after having been captured by the palefaces and lived with them for years, the chief called in a convention of the tribe to celebrate the return of their son. On the occasion, the son was to report about what he had experienced with the aliens. So he described all the peculiarities he had witnessed as to all kinds of behavior, of clothing, cooking, sex, etc. He concluded with a real highlight as to the palefaces' oddities. "You know what they believe?? They think the earth revolves around the sun!!" The tribespeople fell on their backs in laughter.

Centuries earlier and the palefaces still on their own continent, to utter the same weird idea was not a laughing matter. You could be severely punished for it.

Planetary movement is the most famous (or notorious, rather) example in science where the majority had it wrong and eccentrics had it right. Even though Aristarchos had it right in the 3rd century BC by seeing the sun at the center of the planetary system, Ptolemy's complicated system of planetary movements around the earth eventually dominated thinking in astronomy for about 13 centuries, with Copernicus being the first to get it right two milennia after Aristarchos. The story is not quite comparable to modern ones because for many centuries theologians rather than scientists had the last word in scientific matters due to the sheer political power of the Catholic Church[2]. A parallel story actually *did* happen in modern times however when in Stalin's Soviet Union Marxist ideologues (what they had concluded from Marx' teachings they considered to be science) rejected modern genetics and

---

[2] The church condemned Galileo for propagating Copernicus' sacrilegious heliocentric idea to life imprisonment (later commuted to house arrest) and took some four centuries (until 1993 to be exact) to rehabilitate him.

ruthlessly enforced the upholding of the false dogma of acquired traits being encoded in the genetic material (with disastrous consequences for the Soviet agronomy).

The heritability of intelligence – a scientific zombie still going strong

In astronomy, today a scientific field based on physics, science eventually trumped ideology and today no one doubts the heliocentric view. But in science, where one might think that rational thought should finally prevail, things need not happen like that. And it did not happen in a field where, had science trumped prejudice, disastrous social and political consequences could have been avoided: the controversial issue about the heritability of intelligence. Here the opposite happened: prejudice not only trumped scientific reason in the past, but still does so today.

How could something like this happen in a debate deemed to be a scientific one? The treatment of the issue about the heritability of intelligence and other mental traits as it happened in science is one of the crassest and most depressing examples of nonscientific aspects being so dominant as to preclude a consensus about a genuinely scientific problem. More depressing still: the consensus could have been reached by some fundamental and rather simple considerations.

Ever since Francis Galton described the regression effect and mistook it for a law of heredity[3] around the end of the 19th century, scientists have believed that intelligence is heritable to a substantial degree, somewhere between 50 % and 80 %. When quantitative genetics, the statistical tool for determining the heritability of quantitative traits, had been developed in the 1930s, psychologists applied it to intelligence and still do so today, against the warnings of one of the experts in the field, Douglas Scott Falconer[4] (1989). They ended up with heritability estimates between 10 % and 90 %. They also ignored the warnings

---

[3] Galton falsely derived heritability from parent-child regression while this regression is actually affected both by genetic and environmental factors. Regression is not a biological but a statistical law, applicable to all kinds of correlating variables.

[4] Quantitative genetics was developed in order to predict breeding success in plants and animals. The preconditions for its use for human mental traits cannot be satisfied due to the lack of experimental control.

implied in these results; instead they simply took the average and proclaimed intelligence to be 50 % heritable (Neisser et al., 1916). The warning: the large differences in heritability estimates between populations found cannot possibly be genetically caused in a genetically very homogeneous population like homo sapiens. A short look at the definition of heritability in quantitative genetics would have revealed that the differences must primarily be attributed to differences in environmental variation between the populations.[5] As a consequence of all this, there simply cannot be a general value for the heritability of intelligence and, more importantly, a heritability coefficient practically cannot tell anything about the degree to which intelligence can be increased by social intervention (for a more detailed explication see Velden, 2014). And as a consequence of this, it makes little if any sense to do heritability estimates for intelligence.

Molecular genetic studies, which had for a long time been hoped to help clarify things, have not found any substantial contributions of heredity to intelligence.[6] Yet still we are being bombarded by psychologists with proclamations of a high heritability of intelligence (e.g. Plomin, 2018). The evidence lacking, we are being told that it will show up in the future, a strategy which, as we will see, also characterizes the field which is the topic of this book, the attempts to create human-like computers.

As mentioned, research on the heritability of intelligence did not happen on a scientific playground but had huge social and political implications. The most disastrous of the effects of the belief in a high heritability of intelligence were the ones on immigration restrictions for certain countries, on compulsory sterilisation of people deemed feebleminded (a vague, unscientific "diagnosis"), and restrictions on access to higher education for students deemed not to be intellectually fit for it.

---

5   In the simplest case, which can be taken for explanatory purposes, the heritability coefficient (h^2) is defined as $h^2 = \frac{V_g}{V_g + V_e}$, where V_g is the part of the total variation due to genetic differences, and V_e is the one due to environmental differences. V_g being constant in a genetically homogeneous population like homo sapiens, differences in h^2 can only be explained by differences in V_e.

6   A huge (80 000 children and adults) so-called genome-wide association study showed that at maximum 4.8 % of the variance in intelligence may be explained by genetic differences (Editorial, 2017).

There is probably no field of science as contaminated by ideology and prejudice as the one about the heritability of intelligence. This is best documented by a group of psychologists advocating a high heritability of intelligence and a genetically based intellectual inferiority of coloured people: Hans-Jürgen Eysenck, Richard Lynn, Philippe Rushton, and Arthur Jensen, all of them having been supported by the "Pioneer Fund," a racist organisation as documented in great detail by William Tucker (2002).

A further highly detrimental effect of the ongoing propaganda about a high heritability of intelligence is the creation of a whole new branch of science, "Sociogenomics," in the wake of ever cheaper and simpler methods of DNA analysis. It tries to find a genetic basis of all kinds of psychological traits and functions, intelligence among them. It is based on molecular genetics and totally ignores the fact, based on quantitative genetics, that the heritability of mental traits is largely a function of environmental variation (see above). Contrary to permanent claims about molecular genetic indicators for it, most forcefully proclaimed by Robert Plomin, no valid data supporting the heritability of mental traits exist. So even in modern science, where no dogma is enforced by an outside authority, unscientific ideas may be proclaimed with no end in sight. As will be seen, such aberration may be rare, but is by no means unique.

Behaviorism – a mindless psychology

A science robbing itself of its own subject matter, i.e. intentionally ignoring what it had initially planned to study, is unique in the history of science. It happened, though, in psychology with behaviorism being widely espoused. For nearly half a century the idea was seen as a really good one by the majority of psychologists.

Behaviorism, the ideology dominating psychology for more than forty years, was based on the idea that psychology should proceed in the manner of a natural science, i.e. be based on objective observation. The idea was born under the impression of spectacular advances in physics, chemistry, and biology in the late19th and early 20th century. The problem with this idea: human psychological processes can only be observed directly by self observation (introspection), a method producing only subjective and thus quite unreliable data. Watson's (1913) solution: kick all those psychological processes out of psychology. The problem with that idea: you now have a psychology that does not study psychological processes. But psychologists operating under the ideology that psychology must be a natural science were undeterred by the paradox and started the program of a psychology exclusively based on what is objectively observable in animals and humans: behavior. The most prominent among them were, after Watson had to leave academia in 1920, Clark Hull and Burrhus Skinner. Let us have a look at their creations under the ideology.

Hull in all earnest undertook it to explain and predict human behavior on the basis of mathematically describing behavioral elements shown by rats in a maze. Small as those behavioral elements may have been, it took a lot of mathematics to describe them (In Hull's central opus *Principles of Behavior* we find a seven pages long list of mathematical symbols used). All those mathematics impressed many at the time, mistaking the use of mathematics for the essence of what constitutes a science. Some psychologists still think so today. It impressed Hull himself so much that he found that "[…] there is good reason to hope that the behavioral sciences will presently display a development comparable to that manifested by the physical sciences in the age of Copernicus, Galileo and Newton" (Hull, 1943, p. 400).

Still more outlandish than Hull's program (if that is at all possible) was the attempt by Skinner, one of the most popular scientists in America at his time, to explain language. Everything mental prohibited by the behaviorist ideology, he tried to explain language, a means obviously developed in order to transmit meaning, without referring to the concept of meaning and by just using the mechanisms of reinforcement ("instrumental conditioning" in his case), i.e. the reinforcement of the emissions of sounds produced when speaking. Ideology outdoing reason at that time made him not recognize the sheer absurdity of his undertaking and led to a hopeless debate with Noam Chomsky who just stated the obvious: language is about meaning, not just the production of sounds.

Needless to say that hardly anyone today refers to the approach to mental phenomena as it was advocated by psychology's Galileos and Newtons. Hardly anyone, except for a small group of psychologists, the Gestalt Psychologists[7] ("Gestaltists") and Sigmund Koch realized that a mindset, an ideology,

---

[7] After the misdevelopment had occurred and the damage to science had been done, the Gestaltists proclaimed the fact, actually a triviality, that human mental processes must be studied in a meaningful context, a human context.

not scientific reason, had come to dominate the field of psychology, which for me as a psychologist creates a feeling of embarrassment in retrospect about the situation my field had brought upon itself. As Hull's musings about the behaviorists' status in the history of science tells us, the belief in the behaviorist ideology had become a state bordering on pathology, the state of a delusion. Thousands of scientists around the world were afflicted by the condition for nearly half a century. And, more lamentably, generations of students had to learn the oddities produced by the self-proclaimed scientific giants. "Madness of our time" Weizenbaum would probably have diagnosed, had he been a psychologist at the time.

It took some time (more than 40 years) for the psychological scientific community to recognize that a psychology without psychological processes was not such a good idea after all, and it was eventually (around the end of the 1950s) abandoned. The analysis of what had gone wrong and why did not go very deep, however. For the problems with psychology as a science are of a deeper nature than that of acknowledging that mental functions must be the subject matter of psychology. They lie with the idea, upheld today by nearly all psychological scientists, that psychology must *be* a natural science and not just at times proceed like one. This very notion of psychology being a natural science is refuted by Koch's *Psychology in Human Context* (1999), one that draws insights not only from the natural sciences but also from the fields originally dealing with the human psyche, the humanities and, not to forget, from common sense.

As Koch has argued on a broad social-philosophical-methodological level and as I have documented on the basis of the failures of several ambitious psychological projects, some of them quite spectacular (Velden, 2010, 2016), psychology restricted to natural scientific proceedings simply does not work. According to Koch, it is "the institutionalisation of a delusion." Most psychologists working in academia (not the ones working in applied fields) today remain afflicted by this institutionalized condition, however.

Psychosomatics – an etiology lost and regained

As we can glean from the writings of Hippocrates and Galenus, the idea that organic diseases may be affected or even caused by psychological factors was firmly established in antiquity. But in the second half of the 19th century, the century of the great breakthroughs in science and medicine, the idea had practically disappeard from the medical sciences. What had happened? Had the idea been forgotten? Paradoxically it was actively pushed out of the medical sciences due to the dogma dominating medicine at that time, namely that diseases can only be caused organically and that the idea of psychological causation is an unscientific and irrational, if not a mythical one. The dogma of exclusive organic causation had mainly developed due to two great discoveries in the 19th century: the discovery of microbes by Louis Pasteur and Rudolph Virchow's observation that organic pathology is always basically a pathology of cells. The discoveries with their huge beneficial effects had, as is often the case in science, a retarding side effect, namely the false conclusion that *all* pathological symptoms, the psychological ones included, must have an organic cause. To reestablish the psychosomatic idea in medicine took a whole movement, one in which Sigmund Freud played a crucial role by presenting a theory about psychological causation of physical symptoms. His theory was generally accepted in medicine (mind that Freud, before becoming the famous psychologist we know, was an internationally renowned neurologist). Freud proposed that repressed (and thus unconscious) mental contents, mostly those of a traumatic nature, can convert into bodily symptoms like, for example, the inability to walk. Symptoms like this were called "hysterical" symptoms at the time. Today, honoring Freud's theory, they are called conversion symptoms. The acceptance of Freud's theory in medicine sufficed to reinstate the idea of psychological causation even though the hysterical symptoms were untypical and rare compared to all those diseases seen as psychosomatic today, like the coronary or peptic diseases.

The belief in exclusive organic causation as it dominated 19th century medicine was actually an irrational one, a dogma. Its irrationality can be seen from the fact that mechanisms of psychological causation are plausible, simple in principle, and had been known, or at least suspected, all along. It is in no way irrational to assume that psychologically caused bodily malfunctions can eventually lead to organic diseases. High blood pressure, which may contribute to atherosclerosis of coronary arteries and eventually to cardiac infarction, may be caused psychologically, as can be the extreme production of peptic substances which can lead to damage to the lining of the stomach or the small intestine and ultimately to ulcers. Psychological states at times having massive effects on bodily functions has been known as far back as antiquity. In the rejection of the psychosomatic idea in 19th century medicine, we once again encounter the overriding power of belief making us, scientists included, not see the obvious.

Mirror neurons – a debacle caused by hype

Hype, even more so than dogma, may make science not just blind for the obvious but make it lose contact with reality altogether, quite as what happens in delusion. Subjectively felt as euphoria, yet not in any way as an unscientific motivation, it may cause whole fields of science to march into blind alleys. This is what recently happened in neuropsychology with the discovery of so-called "mirror neurons". They are neurons in the premotor cortex of macaque monkeys (and most probably of humans, too) activated when the animal not only executes an action (which was well-known before) but also when it *observes* other animals executing the same action. The psychologist V. R. Ramachandran, always bent on proclaiming sensational scientific breakthroughs, compared the impact of the detection of mirror neurons for science to the one the discovery of the structure of DNA had for biology, reminding us of the above-mentioned claim by Clark Hull about the status of behaviorism in the history of science. Mirror neurons would help explain "a host of mental abilities", even, as soon would be postulated, practically all of them, including risk assessment, sexual orientation, political attitudes, business leadership or, quite specifically, the misattribution of anger in the music of avant-garde jazz saxophonists. Obviously, the mirror neuron hype had finally gone into overdrive. Yet, as has often happened with new trends in psychology, after some 20 years (the mirror neurons were detected in the early 1990s) and thousands of publications, the hype subsided after Hickok (2014) showed that basic assumptions about the function of mirror neurons, like their alleged role in understanding the actions of others, or even in understanding abstract concepts and ideas, did not hold. As it turned out mirror neurons play a minor role at best in complex mental processes in general and in abstract ones in particular.

One among the many hopes the mirror neuron hype had prompted was to find a neurophysiological explanation for autism, the resulting speculations even dominating autism research for a while. Autism being a frequent and serious condition, claiming much attention in the medical sciences, makes the hype-produced mirror neuron debacle, i.e. the fact of the neurons not explaining anything psychological, all the more an embarassing one for neuropsychology.

Localization of mental traits – phrenology's long goodby

One of the most common yet misleading beliefs about the mind-body relation is to think that mental traits or functions may be localized somewhere in the brain. It probably dates back to the time when it became common knowledge that the brain is the organ of the mind. With this idea generally accepted, the question probably soon arose as to *where* in the brain all those psychological traits we have terms for do reside. Such traits like intelligence, strength of charater, or consciousness are called "constructs" or "hypothetical constructs" in psychology. They are abstractions and do not denote things with a reality of their own, like atoms, microbes, or trees. The conceptual step (or misstep, rather) to see these abstractions as having a reality of their own, as being some kind of *thing*, is called "reification". Reification is very common but quite misleading and nearly inevitably leads to the eqally common and misleading notion that those "things" must be located somewhere in the brain. The consensus among modern scientists that mental traits and processes must have a biological substratum, i.e. that they do not exist without a material basis, makes the idea appear quite plausible, as did Franz Joseph Gall's 19th century phrenology to many scientists of his time. Gall presented a complete map telling us where the diverse mental traits can be found in the cerebral cortex. The notion that more or less of a mental trait requires more or less underlying brain tissue and that thus a highly manifest trait would create a bump on the skull, made him think that the character of a person may be diagnosed by measuring the form of the skull. The whole idea is rightfully ridiculed as "bumpology" by many today. The localization idea is not, however.

The idea has repeatedly been boosted by technological advances, starting with the electroencephalogram (EEG, recording of electrical potentials on the skull) nearly a hundred years ago, to today's "brain scanners", the most sophisticated of them being the functional magnetic resonance imaging (fMRI, recording of blood flow in restricted brain areas). The fMRI technology, used in the most ambitious endeavors to localize mental functions, was developed as a medical instrument used to diagnose brain dysfunctions, like a reduced blood flow (and thus reduced activity) in a certain area of the brain. In neuropsychology, however, neurally intact subjects are given certain mental tasks and the machine shows the pattern of brain activity concomitant with the execution.

Among the many technical-methodological problems with the procedure (see Uttal, 2001), one fundamental problem stands out: we do not really know *what* to localize. It is psychology's notorious definitional problems we are dealing with here. As Uttal puts it:

> "There is no historical trend, no evolutionary sequence, no convergence toward a taxonomy of mental components" (p. 145), so that, "the ultimate nature of mind – whether it can be decomposed or must be dealt with as a unit – remains one of the important, yet all-too-rarey considered conceptual conundrums in contemporary scientific psychology" (p.148).

As a consequence, there are practically no successful replications in the field and, "Almost any replication came up with different results that differed either somewhat or completely from those of the original study" (Uttal, 2001, p. 200). And as Harpaz (1999) found: "The data about replication of cognitive imagery between different studies and different research groups is extremely sparse, and the sparse data is virtually all negative" (p. 7).

Most telling about the sad state of the localization debate is the fact that we are still ignorant about the crudest localization, the one about functional differences between the two hemispheres. Ever since Paul Broca in the 19th century discovered the motor speech area, a cortical structure in the left hemisphere (in right-handed people), it was known that the two hemispheres of the cerebral cortex, which at first sight look quite alike, have partly different functions. In the 1960's a new treatment for epilepsy, the cutting of the nerve fibres which connect the two hemispheres[8], appeared to open the possibility to study hemispheric functional specialization in detail, or so it seemed[9]. The treatment was soon abandoned and the restricted access to "split brain" patients made psychologists develop methods to study brain

---

[8] It was performed in order to prevent the spreading of the pathological brain activity from one hemishere to the other.

[9] The procedure allowed presenting visual stimuli in a way that they were perceived by one hemisphere but not by the other.

asymmetry in neurologically intact people.[10] A whole branch of science developed, the study of "hemispheric specialization," the results of which were not rarely popularized, alleged differences between men and women in particular. But after decades of research and thousands of experiments no clear picture emerged. Robert Efron (1990) described the state of the field as chaotic, both as to the functions to be specialized and the factors affecting specialization. He suggested, unsuccessfully, to abandon the field. Of the main and much popularized result that the left brain is more logical, verbal and detail oriented while the right brain is more creative and passionate, spacial and holistic, Hickok (2014) writes that it is "[…] a bit of a popular myth, the twentieth century's equivalent of phrenology" (p. 59).

Let it be stressed that the critique of the localization project does not imply that mental functions cannot be localized in general (see below). The idea behind the project is to find neural correlates like consciousness, insight, or memory, i.e. of psychological *constructs* with the ultimate goal of finding the physical (neural) basis of the psyche as it is described by psychology. It must be stressed that the terms for all these functions, traits and states are constructs. The idea behind the localization project, on the other hand, is based on the naive and simplistic notion that the psyche may be seen as composed of psychological elements, a notion shown to be untenable, even irrational by the Gestalt Psychologists. As a consequence of this flawed notion there is, as we have seen, not anything like a systematic nomenclature of psychological constructs, and as a consequence of *that* the localization project is in a chaotic state and will remain so if it is not abandoned altogether. *Concrete* mental functions (constructs are abstractions) may very well be localized, however. Take, for example, Adrian Owen's (2017) groundbreaking proof that neurological patients deemed to be unconscious (being in a "vegetative state") may very well be completely conscious. When he instructed them to imagine playing tennis they showed the same localized brain activity pattern as did neurally intact persons receiving the same instruction. To imagine playing tennis is not a psychological construct. It is a *concrete* mental func-

---

10   Different stimulus material was presented to the two ears simultaneously and the functional assymmetry was studied via right or left ear preference for one or the other material.

tion, resulting in a specific brain activity pattern as does any other concrete mental function.

The failure of the localization project notwithstanding, the belief in the localizability of psychological functions is unbroken (see the attempts to localize consciousness below), a highly disconcerting situation. Even though lacking any empirical evidence, our intuitive tendency to find the site of everything psychological we have a word for, combined with the tremendous prestige of the technological machinery used to find that site, dominates the field. Paradoxically the machine enhances the prestige of the field, even though the use of this very machinery has shown the whole localization idea to be a flawed one.

The complete failure of those costly attempts to find a neural basis of mental constructs by trying to localize them, demonstrates that that basis is as obscure as it ever was and, considering the undiminished taxonomical problems plaguing psychology after centuries of research, will ever be.

Many more examples of belief overriding reason, and thus making whole fields of science go awry, may be given. But the ones presented may suffice to convey the fact that the phenomenon is not uncommon. The causes for this are manifold. Tradition, thinking in routines, reign of "authorities" in a field, prejudices, intuition (in case of a counterintuitive reality), religious or ideological beliefs, social pressure, financial interests, hype created by skillful propaganda, etc., etc. may make us want to believe rather than know. All these powerful influencing factors may result in a mental state bordering on the pathological, a state in which the attention is totally captivated by the dominating trend and even the most obvious alternatives are being ignored. As has often been observed in mental history, it is extremely difficult, at times even impossible, to get out of this mental state because it is simply not recognized as such. In order not to fall prey to the above influencing factors and, even more so if already under their influence, it is essential to acknowledge not just that this mental state can happen to a person but also that it is quite common even among reason-oriented people such as scientists.

As to the topic of this book, the belief in machines eventually becoming or even already having become human in that they for instance allegedly show consciousness or culture, we are in my view dealing with the mental state described above, a state Weizenbaum diagnosed as a pathological one some 50 years ago. He even went further than attesting madness merely to some scientists by extending the diagnosis to "our time," i.e. to the majority of people today. He may have hoped that the phenomenon would ultimately disappear by virtue of eventual insight on the part of the scientific community, but the opposite has happened. Today nearly everybody dealing with the issue thinks that he *knows* that machines can be like humans. Those, who are somewhat sceptical about the present state of affairs, feel nevertheless mostly compelled to stress that the machines are *not yet* quite human. Under the influence of the ubiquitous proclamation of new breakthroughs in AI research, hardly anyone dares to say that there will *never* be human-like machines, as did Weizenbaum in 1976. Without any hope that the scientific community will eventually have the insight all by itself, the laborious approach must be taken, to show in detail why for fundamental, rational, scientific reasons the machine-as-human idea is a flawed one, that machines cannot be human-like on a-priori grounds. Counter to the consensus among AI researchers not to use the word "impossible," it must be made clear that its use is at times quite inevitable.

As mentioned, that laborious approach means having to paint a realistic picture of what psychological processes actually are, one that is missing in a debate characterized by simplistic ideas about those processes. They are ideas developed *under the provision* that those processes can eventually be made to run on computers.

But let us first have a look at some peculiar characteristics of the computers-as-humans AI field. They may make us doubt that we are here dealing with science in the first place.

AI peculiarities

The field may at first sight look like one scientific field among others, actually one closely connected to the prestigious STEM (science, technology, engineering, mathematics) group. But at a somewhat closer look some characteristics are revealed which are untypical for STEM. Quite like psychology, the field is one of its own in that it has developed its own procedural standards despite appearing to be part of STEM from the outside.

AI – a misnomer

The first and the most consequential of those peculiarities is how the field calls itself. We have become used to the term Artificial Intelligence knowing quite well that the field merely deals with all kinds of information processing devices, most of which we would not think of comparing, or even seeing in analogy to, human information processing, like all the machines in industy, administration or medicine operated by computers. But the use of the term Artificial Intelligence becomes problematic if, as is the topic of this book, it denotes the endeavor to emulate human information processing, i.e. mental functions like, for example, intelligence.[11] The very use of the term Artificial Intelligence in this context implies claiming that, by definition, the information processing in a machine is comparable to the one happening in humans. It is kind of a taxonomic trick used to insinuate that comparability, a comparability which ignores the fundamental difference between computers and humans.

---

11   The term intelligence, once it became used in a psychological scientific context, applies to humans, of course. The first instrument for measuring it was constructed in order to assess the mental capacity of pupils.

Take the example of playing chess. The beating of a master in chess (and then in Go) by a computer was celebrated in the AI community as a great leap forward toward the creation of intelligent machines, thought by some even to be able to surpass humans with respect to that faculty in the not too distant future. But does it make sense to call a machine that can masterfully play chess but can do little, if anything else, intelligent? There exists no known mental structure or state in humans comparable to that. But even if we strained our imagination and thought of a human capable of playing chess only, we would certainly not call him intelligent. We would rather speak of a serious pathological condition, characterized by extreme mental deficiency, quite the opposite of what it is claimed to be in the AI community: a step toward outsmarting humans. Implicit in the use of the term Artificial Intelligence is the highly problematic assumption that what computers do is quantitatively comparable to human intelligence, i.e. that they can be assigned a value on a scale defined for humans. But there is a *qualitative* difference between what goes on in computers and what goes on in humans. A computer may, in an extremely restricted domain like chess, behave like a human. But human intelligence relates to all aspects of human life, meaning that in contrast to computers they show intelligence in an infinite number of domains and circumstances. The fact that in extremely restricted fields computers do behave like humans does not allow us to consider them to any, even a most minimal degree, as human-like. And even in an extremely restricted field like chess, computers merely *behave* like humans. The processes underlying the behavior of humans and that of computers when playing the game, however, are totally different. To say computers are human-like because they play chess, a behavior shown by humans too, is like saying that they are human-like because they can compute, something humans can do too. Human-likeness is not a matter of behavior as such, but of the functions underlying and regulating it. As we recall, to intentionally ignore that difference was the main flaw of behaviorism and the main cause of its demise.

Misleading use of analogies – raison d'être of the field

The use of analogies,[12] like speaking of intelligence in machines, appears to be the main scheme by which AI research insinuates that it can produce human-like characteristics in machines.

The analogy which prompted Weizenbaum to sense madness, was the term socialization applied to machines with the understanding that they could be modified by their experiences with their worlds. Analogies are so particularly apt at associating machine functions with human mental functions because *some* similarity in function between computers and humans, albeit extremely weak ones, can always be found. The above analogy, a functional similarity between computers and humans in that they can both be modified by their environment, is such an extremely weak one and Weizenbaum should not have consented to its use because the use alone makes us think of a man-machine similarity, which, in a meaningful sense, simply does not exist. The socialization of a machine, even though formally possible (having been modified by the environment), is nonsense when compared to a meaningful use of the word socialization. Drawing an anology between man and machine by using the term socialization for both of them is a decisive step toward misunderstanding the nature of human socialization. The analogy is so weak that it can just as well be applied to the similarity between humans and all kinds of dead matter. A rock, too, can be modified by its "experience" with its world, for example if a car runs over it. Interestingly Weizenbaum himself used the term *experience* in the context, which is inappropriate because in a meaningful way a machine can not, as little as a rock can, experience anything. Needless to say that both, the computer as well as the rock, lack the neural-humoral structures and the mental-emotional biography necessary for that. I state the obvious knowing quite well that, except for some panpsychists, no one may imagine experiences in rocks,

---

12   In biology, where the term is defined most clearly, an analogy means a similarity in function, yet not in evolutionary origin. As an example, the wings of birds and insects are functionally identical in that they are both used for flying. But evolutionarily the wings of birds are modified skeletal forelimbs, while those of the insects are extensions of the body wall. So the analogy is misleading as to evolution, quite like the term intelligent, when used for describing the behavior of a machine, is misleading with respect to what happens in man and machine.

while quite a few in the AI community have no problems with computers having experiences.

Even weak analogies may be quite plausible though, like, for example, the very common one where the function of storing and retaining information is seen as being performed by both humans and computers. But it is quite misleading to think that the machine is in any way human-like just because it can store information. The actual working of their memory is totally different for man and machine, the main difference being that human memory is quite idiosyncratic, meaning that in contrast to computers there are not two humans on earth who memorize things in an identical way, even when such rather simple things as numbers are to be stored (see below).

Analogous functions are, of course, not identical functions. To draw an analogy between the function of a machine and a human and then name the machine function after the human one is just a semantic trick (even though a quite effective one) to create the impression that machines can be human-like. Once the analogy starts to dominate our view of two different behaviors, the dissimilarities begin to move into the background of our consciousness or may even disappear altogether. At that point we have no problem, for example, with regarding robots, which just shove a little ball around, as playing football (see below). The AI community has been hugely successful in putting man-machine analogies into the foreground and thus pushing the dissimilarities out of focus or out of our consciousness altogether. It may even happen that an analogy is being used even though it is unclear what the man-machine similarity is supposed to be, as is the case with the quite popular idea of computers being conscious. The idea of machines having consciousness, absurd as it may be, has attracted much attention and caused much discussion. The discussion alone, irrespective of what came out of it (nothing substantial as we will see) has very much favored the impression of a human likeness of computers. It will therefore be treated in quite some detail below.

Predictions masquerading as facts

Another peculiarity of the machine-as-human movement is that it mostly does not deal with established facts but with predictions, promises and announcements, often creating the impression that the announced feats of computers have already been accomplished as, for example, in the case of self-driving cars[13] of which a broad public has been made to believe that they do already exist.

Recently I saw some robots on TV shoving a little ball around. They were "playing football." Typically the commentator asked: "Will they soon play like Ronaldo?" He did not answer the question but left it to the viewers to think that computers might eventually play football Ronaldo-style. Not much later I read that in 2030 a robotic football team will beat human champion league teams. The year 2030 was well chosen. Nobody will remember the prediction at that date. For it to become true is about as probable as an explosion in a kitchen creating a dish of Spaghetti Carbonara.

The most famous overblown prediction is the capacity of HAL, the computer in Stanley Kubrick's *2001 – A Space Odyssey*. HAL cannot only understand human language, it can even read it from the lips. Its understanding of humans is such that it even tries to play on the astronaut's compassion in order to prevent him from turning the computer off. Today, more than half a century after the release of the film, there is still no computer capable of reading language from the lips, even not one to understand human language in the first place. And certainly not one so human-like as to play on human emotions in order to attain a goal. So even our best thinking director, as one critic characterized Kubrick, was deluded by the deus-ex-machina status of the computer.

The standard for grandiose predictions was set long ago (in 1970) by the futurist and "father of artificial intelligence", Marvin Minsky, when he predicted that between 1973 and 1978 "[…] we will have a machine with the general intelligence of an average human being. I mean a machine that will be able to read Shakespeare, grease a car, play office politics, tell a joke, have a fight. At that point the machine will begin to educate itself with a phantastic speed. In a few months it will be at genius level, and a few months after that its powers will be incalculable." (Adee, 2016).

Decades ago the strategy to sell announcements as facts failed and the first "AI winter" befell the field, freezing the means for all those grandiose projects about human-like machines because scientists outside AI had started to become sceptical about what computers can really achieve. Amazingly the strategy is in

---

13   Ones driving without a human operator and in today's normal (unprepared) traffic environment.

full swing again today and the predictions are still as outlandish as was Marvin Minsky's above one. A second AI winter may come (see "the deep-learning bubble" below), but the firm belief in the almightiness of the computer, popular as it is among most scientists, will probably help to overcome it too.

A ghost in everything

A well-known psychological mechanism favoring the belief in the existence of human-like machines is anthropomorphization, our tendency to project human properties onto all kinds of objects, living or dead ones. No owner of an animal, foremost a dog or a cat, doubts for a moment that his animal partner has thoughts and emotions similar if not identical to his own. Friends of plants not rarely feel they have a kind of soul, such that they may talk to them, or hear a tree talk when the wind makes it whistle. It need not even be living matter for it to be seen as having soul-like qualities. Mountains, rivers, rock formations, etc. have typically been seen as not only hosting gods or spirits but also as *being* ones. The river Rhine became "Father Rhine" in the German romantic mythology of the 19th century. As a response the poet Heinrich Heine made him complain about the nationalistic fuss created about him. In the film *Cast Away*, in which Tom Hanks is stranded on an uninhabited island on which he is to live for years, he paints a crude face on a ball and calls it "Wilson". He develops a close emotional relation to his companion, often talking to him. One of the most moving scenes of the film is the one where he falls into despair fearing that Wilson has disappeared.

So it is an almost automatic reaction to feel like talking to a person when interacting with a computer, even if what the computer emits does not make much sense. More than 50 years ago, Joseph Weizenbaum created a computer program that did "psychotherapy", the patient, a human, and the therapist, a computer, communicating in written form. It was in no way intended by Weizenbaum for the computer to really do psychotherapy because, obviously, the machine could not understand what the problem of the patient was. The program eventually became well known under the name ELIZA, but not because people found it funny as Weizenbaum had intended, but because people actually found themselves helped with their psychological problems by the program. Psychiatrists even found the program should be developed further for actual psychotherapeutic use. Weizenbaum was surprised, even shocked by that development because for him the use of a computer program for psychotherapy was a mad, rather than just a bad idea.

The idea that humans might create human-like machines has always fascinated writers, philosophers, and scientists, but today's popularity of the idea is mostly due to Hollywood picking it up, beginning with *Frankenstein* in 1931, in which a fanatical scientist, Dr. Frankenstein, creates a human-like monster. There being no computers around in 1818, the year Mary Shelley's novel was published, and still not in 1931, the year the film was released, Dr. Frankenstein had to go about it the hard way and compose his monster from parts of corpses, which he manages to animate by some scientific knack. To create a human-like monster, machine, or robot for the screen is quite easy: just take an actor and declare him to be human made, like Boris Karloff, the iconic monster, Yul Brynner, a really menacing robot in *Westworld*, released in 1973, one still topped in his menacing quality by Arnold Schwarzenegger in *The Terminator*, released in 1984. The genre has been perfected to a degree that it is hard to imagine that people having grown up with it do not believe in human-like machines. So in my attempt to show that there cannot possibly be human-like mental functions in computers, Hollywood, with its tremendous capacity for manipulating consciousness, is clearly the most menacing opponent.

The peculiarities of the field as described, like the misleading claim implied in its naming, the perpetual domination of announcements over accomplishments, and the ample use of analogies insinuating identities, may, so I hope, already have cast some doubt on the scientific basis of its claims and even shown that the field is anything but a coherent field of research.

The notion of human-like computers dates back to an analogy that came up right along with the creation of the first electronic computer: "The brain as a computer." Problematic as this notion may be, it has become the most popular and influential one of all, understood by many today in the literal sense of the brain *being* a computer (as to why it is not, see below). The brain being the organ of the mind (Luria's phrase) the notion directly implies that a computer must be human-like. So the tremendous prestige of the computer (largely well-earned) made it become the deus-ex-machina[14] for solving the unsolvable problem of making a machine human-like.

---

14   The deus-ex-machina is a person or thing introduced in fiction or drama to solve an apparently unsolvable difficulty.

Let it be mentioned as an aside that the ever ingenious Elon Musk, founder of Tesla, has pointed to a perfect solution of the man-a-machine problem, even though not one to be achieved in the near future. According to Musk, the brain should be trained to work like a computer instead of the other way around. For him, the computer aficionado, the computer works a lot more efficiently than that outmoded brain of ours (Siemon, 2019). The more machine-like a human, the better. Logically the next step would be to eliminate humans altogether and leave things to computers, a development predicted by others before. This is clearly the most radical method proffered to solve the man-a-machine problem. With mankind absent, this would actually have solved all problems of mankind because an absent mankind cannot have any problems in the first place. Musk's proposal is probably an extension of the aforementioned general tendency in AI research to start with known functions running on computers and finding analogous functions in humans, rather than first describing the human mental functions and then trying to emulate them with computers. The reason for this reverse procedure probably is that a detailed characterization of human mental processes, rarely found in the psychological scientific literature, would immediately reveal the futility of the whole computers-as-humans project. That detailed (yet still largely incomplete) characterization will be given below.

Human-like computers – a belief system

From a rational, scientific standpoint the case against the computers-like-humans notion is actually a rather simple one. It would suffice to point to the trivial fact that all preconditions for the emergence of mental capacities such as a brain, a human body, or a human environment and biography, are missing in computers. It would also suffice to point out that all claims to have already created human mental functions in computers are but based on weak analogies denoting merely distant similarities, and not identities of functions as claimed.

But the problem with the outlandish AI claims, a problem which makes you rather helpless as a scientist, is that we are not dealing with scientific reasoning but with a belief system here, an overarching mindset deeply entrenched in the unconscious of many by decades of highly popularized claims, helped forcefully and effectively by a vast amount of popular media coverage, including Hollywood, specialized in conveying ideas by powerful emotional visualizations.

For more than half a century now, children, who often do not so clearly distinguish between fact and fiction, have learned that robots can be indistinguishable from humans, that they can *be* humans. They never learned that there may be a problem with the idea. Growing up they have come to think that computers being like humans is common knowledge, a fact.[15] As a result it never comes to their mind that they are dealing with what is but belief.

---

[15] Watching dozens or hundreds of movies in which a machine looks, acts, thinks and feels like a human makes the difference between being created by humans and created by natural reproduction appear as irrelevant as the one between being artificially (in vitro) and naturally reproduced. They are all humans. This is quite ok in fantasy and fiction. In reality, which is where science happens, it is a serious misconception about what "human" means.

Breaking out of a belief system (an ideology, a religion) is an onerous, step-by-step process like the one Sigmund Koch went through when he abandoned his scientific belief, held for decades, that the human psyche should be approached like anything else in the natural world (Koch, 1999).

So, against a strong inclination to just say that the whole man-a-machine idea is an absurd one, period, I have chosen the onerous path and will thus step by step describe the human psyche in its daunting complexity. This will make it evident that it cannot possibly be simulated by a machine. Part of that daunting complexity is the fact that it cannot be analyzed into single, discrete functions as suggested by its presentation in psychological textbooks. As mentioned before, the psychological scientific community itself has not been helpful in conveying that daunting complexity because, ignoring William James' warnings, it itself believes in studying exactly such discrete functions separately, including the interactions between them. As also mentioned before, psychologists have not been very much motivated to present the human psyche in a way that makes the man-a-machine project look naïve, as they wanted to participate in that highly popular project themselves.

As with everything in science, it all starts with defining what we are talking about. In the computers-like-humans AI field that is already the point where the problems begin.

The phenomenology of mental functions – a neglected issue in psychology

The project to create human mental functions, as well as most of the debate about it, suffers from a deep ignorance of the fundamental problems plaguing the definition and characterization of those functions. AI people, and most psychologists too, seem to think that those problems have been solved during the 150 years of psychological scientific inquiry. Yet nothing could be further from the truth.

What are the mental processes as studied by psychology? Where could an AI researcher, intent on emulating them and thus creatig a machine with human characteristics, look up how the experts, the psychologists, define and characterize mental functions and what they have found out about them? He will find them in the curriculum about psychology's basics under the heading "General Psychology". There seems to be agreement among psychologists that the functions generally thought to constitute the human psyche are perception, memory, thinking, language, learning, motivation and emotion. In psychological textbooks we find variations in the headings, such as "Sensation and perception," "Perception and detection," "Memory and cognition," etc. They indicate that there are no strict categories for mental functions, but that interactions among them must be considered in studying them. So, for example, "Sensation and perception" indicates that the results of a perceptual process, called "percepts" in psychology, may be seen as composed of more elementary processes like, for example, a sensation called "loudness" in the process of hearing, or "hue" in the process of perceiving color. The notion, still prominent in psychology today, that the human psyche may be conceptualized as being composed of discrete functions, has been critisized quite early in the history of psychology, most prominently by a movement called "Gestalt Psychology." The Gestaltists even found that forming categories of mental functions hindered more than it helped in understanding the human psyche. As an example, take the perception of a social situation. In addition to the structure of the situation, the way it is perceived depends on one's motivation to understand it, the emotional state (for example of being frustrated) one is in, the experiences one has had with similar situations in the past, which include memory and learning processes, and what one thinks about the situation now, which implies the category of thinking or, as it would be called today, "cognition." So the social perceptual process, even if seemingly as simple as the conversation between just two people, must be thought of as involving the most diverse mental processes, interacting in a qualitatively unique fashion at a certain point or over a period of time.

The one person to have seen the problem with categorizing mental functions first and with great clarity was the American philosopher, physiologist and psychologist William James. He found that the approach to the human psyche a student of psychology has to adopt, i.e. of first studying the diverse mental functions as they are listed in the textbooks, means having to take the second step before the first one. According to James, the first step should involve describing what happens when a person is in a conscious state whatever the situation, a process he called "stream of consciousness" or "stream of thought" (James, 1890). Any further analysis of mental fuctions must start with a description of this process. It must be the subject material of any science of psychology. Trying to describe it, it becomes immediately clear that single, elementary functions cannot be studied independently from all the others. To assume that they can is critically called "elementism" in opposition to the "systems approach" in which a function is always considered in context. Think of what is going on in the mind of a conscious person at any period in time. It is a continuous flow of all kinds of mental processes interacting in a unique way. Some, if not all of our sensory systems supply us with the most diverse percepts, like visual, auditory or olfactory ones. Seeing, hearing, smelling objects being mental events in themselves (perceptions) will evoke memories connected with them, which in turn may prompt us to think about this or that event connected with them. The percepts, memories, or thoughts will, to a smaller or stronger degree, evoke the most diverse emotions. All those percepts, memories, thoughts and emotions may lead to further imaginations of percepts, memories, thoughts and emotions, and so on. This flow of mental events, this "stream of consciousness," is unique for every person because all people have a unique personal history and live in a unique psychological environment, which alone makes for a practically unlimited number of individual mental conditions. And, not to forget, the mental state of each individual is subject to continuous change from one period of time to the next. Seen in this way, there is thus an unlimited number of possible sequences, combinations or interactions of mental events, all of them different for individual people and different at different moments for each single individual. This fundamental psychological situation, a truly unlimited number of possible mental processes occurring both simultaneously and in sequence, must be taken into account when trying to paint a realistic picture of psychology's subject matter.

In order to bring some structure into this confusing situation psychology as a science tried to isolate single mental functions or even elements of functions to describe and study them. Taking physics, where there *are* elements, as its model, it did not pay much attention to James' warnings. So let us look at the fundamental problem psychology is confronted with, the question of whether there are mental elements, a problem evident right at the inception of psychology as an empirical science. It is intimately related to a more fundamental problem plaguing psychology: the fact that the subject matter of psychology, the mental functions, traits and states, like thinking, intelligence, or consciousness, often called "constructs," cannot be observed objectively in the same way as subject matters in the natural sciences.

The problem can best be illustrated by the first attempt to identify mental elements. It happened in psychophysics, the oldest part of experimental psychology, more than 150 years ago.

Psychophysics is a subfield of the psychology of perception which is why the mental elements dealt with are *sensational* elements. Sensations are thought of as one-dimensional, quantitative mental processes, like the experience of loudness of a sound, brightness of an object, or the experience of weightiness a weight exerts on the hand it holds. Ultimately psychophysics was supposed to describe the relation between the size of stimuli (e.g. sound intensity) and the size of the sensations they elecit in an individual exposed to them (experienced loudness) in mathematical form. Unfortunately (one may also say typically), it ran into serious trouble right after its inception.

For critical observers like James (though not for most psychologists of the time who found psychophysics to be a great scientific progress) it was immediately clear that the concept of sensational elements was an oversimplification, even an illusion, meaning that mental functions could not, as one had hoped, be seen as composed of elements. The concept of sensational elements turned out to be totally elusive. The derivation of the most famous mathematical description of the relation between stimulus and sensation magnitudes, still referred to in modern psychology textbooks, the one proposed by the polymath Gustaf Theodor Fechner, illustrates the point. It was based on the critical assumption that the size of a sensation (e.g. loudness) may be conceived of as being composed of single sensational elements or units. These units were the sensational increments corresponding to the barely noticable increments in stimulus magnitudes, for example the increment in weightiness of a weight corresponding to (or caused by) a barely noticable increment in that weight, the so-called difference threshold, for example the increase of the weight from 100 grams to 103 grams, or from 1000 grams to 1030 grams. (We see that the size of a difference threshold depends on the absolute size of the stimulus to be increased). Fechner's assumption was that the experienced increase in weightiness (or loudness, or brightness) of a stimulus was always the same, irrespective of the absolute size of the stimulus to be increased. The assumption appeared plausible according to the motto "a difference threshold is a difference threshold," i.e. you can just notice a difference. But it is in no way plausible if you take a closer look at what it means to experience a difference threshold. That closer look most psychophysicists obviously did not want to take in order not to have to question the reasonableness of the mathematical derivation of Fechner's famous law. The derivation was seen by many as particularly elegant because it was a purely mathematical-deductive one. Elegance or no elegance, James found the derivation nonsensical, even though mathematically correct, because the above assumption that all difference thresholds are experienced in the same way is actually a false one. Nobody can know the size of the experienced sensory difference that corresponds to the difference threshold. In contrast to the difference threshold, which can be measured objectively (in physical terms), the experienced difference corresponding to it is by definition a subjective event and cannot by some methodological trick be translated into an objective measure. All we can do is to speculate about its size, the term indicating that we are on extremely shaky ground here. We are confronted with psychology's fundamental problem, obstinately ignored by most psychologists, namely that psychology deals with constructs, which are hypothetical, not objectively definable entities, like intelligence or, in the present case, the sensation corresponding to the difference threshold. Such a sensation is embedded in the stream of consciousness and therefore affected by the most diverse mental processes. It is an illusion upheld by the psychophysicists of the 19th century and many pychophysicists (and psychologists) today, that if you carefully restrict the environmental conditions, as is typically done in a psychophysical experiment, you have the subject isolated in a way that makes him concentrate on and sense nothing but the sensory dimension you are studying, like, for example, loudness. You can insure that the subject responds to nothing but the stimuli you present because that is all that is given to him. But the experimenter's control is just one of the external environment, not of the internal one, which involves attitudes, prior experiences, memories, emotions, and many things more. These may all affect the sensation studied and thus make it anything but one-dimensional. So if you compare the difference threshold of 3 grams, which you get if you increase from 100 grams, to the one of 30 grams, which you get if you increase from 1000 grams, it may well be that the two corresponding sensational differences are different because the subject may, for example, experience the differences in proportion to the absolute size of the stimuli, meaning that the experienced difference is greater in the latter case than in the former. Thoughts, sensations, feelings, etc. not being separable from each other, form a multidimensional experience from which no simple, one-dimensional sensation may be extracted and then judged by the subject as to its size. As a consequence, all kinds of psychophysical "laws" describing the relation between stimulus and sensation magnitudes have been found. The conditions influencing sensation magnitudes are so manifold and variable, even in the highly restricted situation of the psychophysical laboratory, that no generally binding law can be found in the first place.

I have gone into this in some detail in order to demonstrate to anyone intent on emulating psychological processes with a machine that they are going to face the problem that even on the most elementary psychological level it is unclear what it is that is to be emulated.

Let it be stressed that the problem with the missing elements (and all the problems related to it) will not go away. It's not that psychological elements have not yet been discovered but that, as the Gestaltists have convincingly shown with an abundance of examples, such elements do not and cannot exist.

The biological basis of mental functions

As mentioned, the debate about the possibility of mental functions in machines does not only suffer from a deep ignorance about the nature of mental processes, but also from a naive view of our knowledge about the neural structures underlying those processes. Attempts to create human-like functions in machines often starting with the idea to mimic those structures (an example will be given below), let us look at today's knowledge about those structures.

There is a consensus today among scientists that there must be a physical, largely neural, basis of mental functions. This does not mean, however, that there must be discrete, clearly definable neural structures underlying mental functions. For these functions are, as we have seen, not clearly definable, not even all the same for different people and at different points in time.

Paradoxically an AI researcher looking for such structures may be encouraged when reading psychological textbooks about perception where networks are described that seem to underlie mental functions. But how can that be? How can there be actual neural processes, i.e. not ones speculated about but ones found empirically, underlying psychological ones, even though on the most basic level we do not know what these psychological processes are in the first place? The answer is quite simple: the neural processes presented in those psychological textbooks are not ones underlying psychological processes. So why are they being presented in such detail in psychological textbooks even though the student of psychology cannot learn anything about psychology by studying them? The answer to this question is quite simple, too, even though admittedly somewhat speculative: the presentation is thought to lend prestige to the science of psychology, which in recent decades has tried to masquerade as a biological science, a "brain science." That dubious motive not withstanding, let us have a closer look at what those basic neural processes are and how they relate to psychological ones.

Perceptual processes are always thought to begin at the receptors, cells sensitive to whatever emanates from a perceived object, like sound waves in case of audition or light (electromagnetic waves) in case of vision. In vision, for example, these are the light receptors in the retina of which some are sensitive to light in general and others sensitive to specific bands of wavelengths. It has long been known that even some quite complex processing of visual information happens in the retina, with some of that processing still not fully understood today. But these processes were never seen as neural processes underlying *psychological* perceptual processes, like the formation of percepts, i.e. the conscious realization of an object, of a person, a scene, a landscape, a situation, or whatever. These psychological processes have long been seen as being based on *cortical* neural functions. So when experiments showed that there are cortical cells sensitive to complex visual stimuli, psychologists were alerted, sensing that the neural basis of *psychological* visual processes would soon be revealed. As a consequence, those experiments were immediately presented in psychological textbooks, even though it was by no means clear whether they would indeed eventually lead to an understanding of the neural basis of *psychological* perceptual processes.

The scheme of those experiments was to present visual stimuli to animals (e.g. macaques) and record the ensuing neural activity of single cortical cells in the visual cortex. The complexity of the visual stimuli presented ranged from white or black bars of different lengths and orientations to hands of the animals and eventually even to faces of the animals. The cells were typically called "detectors" because they only responded to ("detected") specific characteristics of the material presented. They were eventually termed "complex cells".[16]

Can those cortical cells (and the still largely unknown neural processes leading up to them and enabling them to do the detecting) be seen as the neural basis of percepts, i.e. of psychological perceptual processes? Clearly, they are a precondition for percepts to be formed. Before being consciously recognized a stimulus must first be analyzed with respect to aspects like form,

---

16 The experiments were first performed by David Hubel and Torsten Wiesel (e.g. Hubel & Wiesel, 1962) who shared the 1980 Nobel Prize for Physiology or Medicine together with Roger Sperry. An identical scheme of neural representation of stimuli exists in other sensory domains, too.

color or movement, processes performed in parallel by the visual system. But the cells resulting from these processings do not detect stimuli in a psychological sense, the one insinuated by the term. Physiologists using the term "detectors" will probably just do so by way of an analogy and for short, not thinking of humunculi, little individuals in the brain doing the detecting. Psychologists, however, should use the term with caution in order to avoid the impression that we are here dealing with psychological processes. But maybe they do not want to avoid the impression in the first place because they like it. It allows them to present these neural processes, not by logic but solely by association, as a neural basis of psychological processes, making psychology appear to be based on neurophysiology and thus posing as a biologically based science.

The strategy to make the complex cells misleadingly appear as the neural basis of psychological processes becomes obvious, for example, if we look at how Goldstein (2014) in his introductory textbook on perception explains how complex cells relate to psychological perceptual processes. To demonstrate the relation, he refers to the phenomenon of "selective adaptation" in complex cells. Take, for example, complex cells specialized to "detect" vertical lines in the sense that they fire if vertical lines are being presented, but not if lines of a different orientation are being presented.

When vertical lines are presented for a longer duration, the firing rate of the corresponding cells decreases: they adapt to the presentation of vertical lines. This effect is called selective adaptation because the effect only occurs in the cells responsive to vertical lines, not cells responsive to lines of other orientations. As a consequence of the adaptation, the threshold for detecting vertical lines is increased, i.e. the sensitivity for such lines is reduced. As a consequence of *that*, the subject needs a higher brightness contrast between the lines and their background in order to still perceive the vertical lines. As Goldstein now sees it, this demonstrates that the detectors play a role in the psychological process of perception. This is obviously so, but why is this self-evident fact specifically mentioned in the context of complex cells, but not in the context of more basic physiological processes, like, for example, ones occurring in the retina, which play a role in the perceptual process too? If you, for example, look at a brightly lit white disk for some 30 seconds without moving your eyes and then look around you will see a dark disk at the center of your visual field, a so-called afterimage. The physical basis of the effect is similar to the above one with the complex cells in that here the receptors in the retina affected by the bright light from the white disc adapt, i.e. fire less and thus create the impression of a dark disc. They have, self evidently, an effect on perception in the psychological sense, but obviously the receptors in the retina are not anything like the neural basis of the psychological process of perception, as little as are the complex cells. No one would specifically mention the trivial fact that retinal processes, for example the ones occurring in dark adaptation, affect perception. But retinal processes obviously cannot be the neural basis underlying perception in the psychological sense, i.e. the conscious realization of some object, and neither can complex cells. But they are situated in the cortex, the structure where we also find the substrate of mental processes in general.[17] The mere fact that the complex cells are situated in the cortex makes it possible to relate them to mental processes on a purely associative yet not logical basis. So why deliberately evoke that misleading impression? As mentioned, aspects other than scientific logic may play a role here.

Anyone intent on emulating human perceptual processes should be well advised not to start at the level of complex cells. All we know about complex cells is that they do exist, but even for relatively simple stimuli, for example a black bar of a specific length, width and orientation, we do not know what the neural processes are which enable the cell to detect that bar. Needless to say that we don't know them for detectors of more complex stimuli like hands or faces.

A neural scheme leading to a cortical cell firing when a face (any face) is presented exists in every person because every person has ample experience with faces in general, as they all share specific characteristics. So we have an area in the brain, the inferotemporal cortex, with cells specific for faces in general, which, if lesioned, leads to an inability to recognize faces, a pathology called prosopagnosia.

As it turned out, there are not just complex cells for faces in general but also ones for specific faces. For a single cell to detect a specific face, ample experience with that specific person is required. If the person is hugely popular, like a filmstar or an American president, we may find detectors for him

---

[17] The complex cells are found in the occipital cortex while the neural structures underlying perception in the psychological sense are situated in the frontal cortex.

in many people and thus there may be, at least for some period of time, a neural scheme leading up to a Meryl Streep detector.

The principle underlying the development of detectors for specific faces – ample experience with those faces leading to them being represented in the brain in the form of specific neural schemes – implies that there must be such schemes specific for groups of people (like vocational groups for whom certain objects or persons are of superior importance) and also ones specific for individual people due to the fact that for every person there are objects or persons relevant to him exclusively, like his dog, friend, or favorite picture on the wall at home. So, beyond the general neural schemes, like the ones for faces, there are group-specific and idiosyncratic ones, i.e. ones for single individuals. Considering the unlimited number of possible individual-personal circumstances, the latter may even constitute the majority of neural schemes in a population.

But even for the universal neural schemes it is hard to establish a binding system of categories, so-called "visual gnostic categories," as the nine proposed by Konorski (1967) like ones for small, manipulable objects, animated objects, faces, emotional facial expressions, etc. Such systems must heavily depend, among other things, on geographic and cultural differences. Think of people living in tropical areas for whom categories referring to snow, ice, or seasonal differences are quite irrelevant, while they are dominant for people living in arctic regions. Still complicating matters is the fact that the individual as well as the general stimulus environment is subject to constant change and so are the corresponding neural representations. So, as a simple example, Konorski's category "handwriting" may disappear because, with the universal use of keyboards for writing, the category will become irrelevant and accordingly the neural representation will disappear.

On a more complex level, the serious problem must be dealt with that the "digital age," the use of the internet in particular, will fundamentally change our routines of learning, memorizing, thinking or interacting socially. The new routines, if they persist long enough, will lead to the loss of old and the creation of new neural representations. Once such neural schemes are formed, the corresponding routines can only be changed with great difficulty. This is alright for routines which must be expected to be efficient in the long run. But many of the routines that have developed in the context of computer and internet use are rather counterproductive with respect to acquiring a solid and comprehensive education, which includes the universal, timeless competences, such as logical thinking, structuring of topics, or the concise verbal expression of thoughts, necessary in an ever-changing world. Such an education is impeded, if not precluded, by, for example, the habitual use of highly selective snippets of information from the internet which are not integrated into a meaningful context. (As to the effects of such newly developed routines and their corresponding neural representations, some of them plain dangerous, see Carr, 2010).

In order to get an impression of how far things have developed with respect to the simulation of neural functions by means of computers, we must step down several levels from the complexity level of functions relevant in education and, for instance, have a look at a quite recent study about the mimicking of neural functions underlying navigation (orienting and moving in space; Savelli & Knierim, 2018; Banino et al., 2018). In that study a "deep learning network"[18] was used for finding a goal in a maze. There were three simple routes of different length to the goal. The shortest one was first blocked and then reopened. Among several deep-learning networks there was one capable of finding out that the shortest route had been unblocked and of then using it as a way to the goal. It had "learned" to use the unblocked route. The successful system allegedly had developed cells similar to ones actually found in mammals, so-called "grid cells," cells that fire if the animal is at any of a set of points that form a hexagonal grid pattern across its environment.

As so often, I feel somewhat uneasy judging this achievement as to the respect it deserves. In the context of the research going on in the field, it is certainly a fine one, worth to be presented in *Nature*. But in the context of human learning, which is ultimately to be mimicked according to the self-set goal of AI research, I doubt that it will lead anywhere.

Note that with the term "deep-learning" we are once again dealing with just an analogy to what occurs in humans. Formally, in the broad sense of responding to a change in the environment, learning does indeed occur. But the behavior of the system is not comparable to anything one would call a learning process in humans, not even to one occuring in a task of utter sim-

---

18   The computer networks are called "deep" because they consist of sequential layers of repeated computational units. Each unit receives input from similar units in the previous layer and sends outputs to those in the next.

plicity like the one presented in the above study.[19] This difference between the learning of machines and humans is reminiscent of a psychological scientific controversy about the nature of learning in the first half of the last century. On one side, the behaviorists, taking as their empirical basis the behavior of rats in a maze, saw learning as an incremental, step-by-step process leading to a goal by means of reinforcement of small behavioral elements. By contrast, the Gestalt Psychologists found that human learning is typically one of problem solving in complex situations. The empirical basis for their ideas about learning in complex situations was provided by Wolfgang Köhler's experiments with chimpanzees (Köhler, 1917). According to the Gestaltists, learning is not a continuous, step-by-step process but one characterized by insight at certain points during the learning process when the elements of the problem solving situation (for example food being outside the cage and not reachable by the animal; two sticks available, each of them too short to reach the food, but connectable to be long enough to reach it) were arranged in the animal's mind (put the two sticks together to reach the food).[20]

In maze-learning, a situation rather untypical for the environment of humans, the reinforcement of behavioral elements may actually occur in humans, too, but the step-by-step approach as we find it in deep-learning, is certainly inadequate for describing typical human learning situations, like learning math or history. So deep-learning may actually be quite helpful at times (it allegedly helped a computer to beat a Go master), but will fail with certainty when trying to simulate human learning, which is charcterized by meaning and insight. (As to the fundamental deficits of deep-learning with respect to creating human mental functions, see Marcus & Davis, 2019)

As shown above, complex cells, i.e. single cells in the visual cortex which function as detectors, are misleadingly presented in psychological textbooks as if they were part of the neural basis of psychological functions. AI scientists may therefore be deluded to take them as a model for computer simula-

---

19   A human would not use a step-by-step process in the situation as does the deep-learning algorithm but would rather proceed by, "insight" (see below).
20   In human learning the subjective experience of the successful mental arrangement became known as the "Aha-Erlebnis" (Aha-experience). It corresponds to "eureka!", the exclamation attributed to Archimedes on discovering a method for determining the purity of gold.

tion because the underlying scheme may appear to be rather simple, the simplicity being suggested by the fact that we are dealing with single cells. But the underlying scheme, the neural connections leading up to these cells and enabling them to do the detecting must be anything but simple. I am saying "must be" because for the great majority of these cells those neural connections are still unknown.

Let me now make the literally long jump (in terms of cortical distances) from complex cells in the occipital lobe to the frontal cortex, where the neural substrate of perception in the psychological sense, the formation of percepts, must exist. About that neural substrate we know nothing. Remember that even the crude (and quite popuar) left hemisphere/right hemisphere localization, the left hemisphere being more logical, verbal, and detail oriented, the right one being more creative and passionate, spatial and holistic, is seen as a popular myth by Hickok. All we know about the neural basis of percepts is that it must be of a daunting complexity, which follows from a well documented fact: the daunting complexity of the mental functions involved in perception. Not two people do perceive the same object in the same way, not to speak of the perception of such complex "objects" as social situations. This idiosyncrasy, or person specificity, sets a-priori limits to emulation, which, after all, depends on rules and laws on the basis of which to proceed. The more we progress following the perceptual transformational processes as they happen on the way from the retina to the frontal cortex, the more idiosyncratic those processes become. On the level of the retina they are largely identical for different persons, as, for example the ones involved in brightness contrast. On the level of the complex cells, there are, as we have seen, already substantial differences between people due to their different visual experiences. On the level of percepts happening in the frontal cortex, idiosyncrasy abounds and rules are largely absent[21], making emulation impossible.

Mental processes, resulting from an uncountable number of learning steps that stretch over decades happen on a physical basis, a substratum that has evolved over billions of years since the first animals appeared on earth.

---

21 Psychologists have elaborated some rules for psychological processes, like the defence mechanisms, but these rules are not such as to imagine a concrete neural basis of them. Freud actually proposed a six neuron scheme for repression, but never elaborated further on it. It was probably just conceived as an illustration rather than a real neural net.

The absurd AI notion that a physical basis could be created out of dead matter (computer components) enabling all these learning processes (which would make it the most complex thing on earth) reminds one of the Book of Genesis where a superhuman being (God) creates ("designs", according to modern creationists) this substratum along with the rest of the world. Theologically there is no problem with that process of creation, assuming that the creator is almighty; but scientifically the idea of humans assuming the role of the Almighty is an absurd one. The simplification involved in the creationist mystique – of a "design" instead of billions of years of evolution – may be unconsciously at the heart of the AI notion that human-like computers can be "created". From the evolutionary view, the one a scientist must take, the concept of creation must be seen as a theological and not as a scientific one. The same applies to creationism in disguise, namely the belief that with the help of the omnipotent computer human-like machines can be created.

After so much time, money, and brain power having been invested and with computers still as stupid as ever (see below), it is time to contemplate the possibility that it is not progress being somewhat slow, but that the whole project of creating humans is a flawed one.

Does "behave like" mean "*be* like"?

Whatever the problems with the approach on the neural level, there still is the behavioral approach, the one taken, for example, when computers were taught to play chess. Humans playing chess (or any other game) does indeed imply a mental function. But can the programming of a computer to play chess tell us anything about the mental processes occurrig in humans when they play chess? It can tell us as little as the program that makes a robot walk can about the sensory-motor functions enabling a human to walk. So did it make sense scientifically to train a computer to play chess in the first place? AI researchers try to make us believe that it did, celebrating the first time a computer beat a chess champion as one great step forward on the way to making computers human-like, even superhuman.[22] Does the fact that computers play chess, or do anything humans also do, make them in any way human-like? An orthodox behaviorist (if there is one left today) would of course like the idea. If in psychology behavior, and behavior only, counts (as John Watson postulated) and the underlying mental functions are not of any interest (as he postulated too, with their study being derisively called "mentalism"), then the behavior of computers and that of humans playing chess must be treated as equal, and the idea may come up that humanness could be approximated by teaching computers more and more behavior shown by humans. Considering the unlimited number of possible human behaviors, the idea is quite unrealistic, however, to say the least. To say "All behavior must be treated as equal, be it the one of a computer or that of a human" sounds reasonable to AI researchers as did "all behavior must be treated as equal irrespective of what is going on in the head of the behaving individual" to John Watson, the creator of behaviorism. As we learn from the history of

---

22   The next pompously celebrated alleged step toward humanness was a computer beating a Go master.

psychology, however, the behaviorist program failed, and there is hardly anyone in psychology left today who programmatically ignores mental processes. Quite to the contrary, modern scientific psychology, which is largely cognitive psychology, tries to study mental processes with the help of the latest technologies, such as the so-called brain scanners. Rightfully Watson's idea of a psychology without mental processes, a psychology without psychology, is deemed nonsensical today, even absurd. But the idea that by teaching machines human-like behavior you may create a human-like mind is all the same not seen as nonsensical and certainly not as absurd today. To the contrary: simply because it involves the use of the almighty computer, that deus-ex-machina, it is seen as particularly sophisticated modern science.

The main cause for AI research pushing toward an unattainable goal, the creation of human-like machines, is thus the fact that the program is seen as a technological one and thus as doable, quite like many other technical or technological programs. The researchers in the field are mostly engineers trained in computer science and mathematics and are as such unaware of the grave conceptual problems that have been plaguing psychology as a science since its inception. The great majority of them have never heard of those problems and thus do not realize that they are trying to simulate something that cannot be defined. Today psychology presents itself as a natural science, at times even as an exact one, and as such does not make psychologists suspect that fundamental problems await them on their way, fundamental here meaning unsolvable. Typically, having dropped the word "impossible" from their vocabulary, they simply cannot (or do not want to) imagine that such problems might exist.

Sadly, there are little if any warnings coming from the psychological scientific community. Today's psychologists (actually the ones working on psychological "fundamentals," not the ones working in applied fields) misleadingly behave as if those problems were just ones of an ancient past and had been solved long ago, making psychology a natural science among others. But as a psychologist I feel compelled to address those fundamental problems and explain why they set close limits to the project of making machines human-like. I will not do this on an abstract level but by examples, the most telling one being the concept of consciousness which is central for understanding human mental processes. But let me first address the metaphor which has affected the debate more than any other: the brain as a computer.

The brain is not a computer[23]

In spite of our extremely limited knowledge about the functioning of the brain and the weakness of the metaphor, the idea of the brain functioning like a computer has been stubbornly upheld till today (it dates from the time of the first electronic computers) by AI researchers as well as by many from other fields, like Stephen Hawking, the physicist.[24] The idea is still so popular that the European Union and the government of the United States finance billion dollar projects aiming to simulate the human brain by computers (Shen, 2013).

It must be doubted, however, that the politicians deciding about those projects were well advised by the predictions about the "digital future" made by AI greats like Ray Kurzweil (2005) who found that "By the mid 2020s it is conservative to conclude that we will have effective models of the whole brain," a prediction based on "IBM [...] creating a highly detailed simulation of about 10 000 cortical neurons [...]" (as a comparison: the brain has about 100 000 000 000 neurons), and that "the technological advance in the 21st century will be in the order of 20 000 years of progress when measured by the rate of progress in 2000 [...]", leaving us wonder how he came to derive that figure.

The notion of the brain working like a computer is at the core of the computer-as-human idea, with Ray Kurzweil, as so often, holding the most extreme view about the function of the brain. To propagate his outlandish ideas without them immediately being recognized as such he resorts to the typical obscuring language as used by the futurists. Take, for example, his

---

23   The first to conceive of the brain as a computer, Walter Pitts, abandoned the idea after studying the neural connections in the frog's eye (Heaven, 2018).

24   Harry Collins (2018) cites an AI researcher saying "I wish Hawking wouldn't write about AI, we don't write about black holes."

declaration that "[…] the mathematical techniques that have evolved in the field of Artificial Intelligence are mathematically very similar to the methods that biology evolved in the form of the neocortex" (Kurzweil, 2010). Well, the semantic problem with biology evolving something (to evolve is an intransitive verb) notwithstanding, I am unable to grasp what "methods that biology evolved in the form of the neocortex" might mean (is the neocortex a form of methods? And what might *that* mean?). But let us, to give Kurzweil the benefit of the doubt, assume that he means that mathematical models have been created to describe the function of the neocortex. The neocortex being the organ of the mind, i.e. the biological basis of the unlimited number of possible mental functions (and thus being the most complex thing on earth), it is hard to imagine mathematical methods describing its functioning while, as we have seen, there is even no valid mathematical description for the relation between stimulus and sensation magnitudes in psychophysics.[25] I think that nonsensical statements like the above one can only be seen as reasonable if the one who proclaims them is in the role of some kind of pope or guru. What, through the fog of his garbled language, Kurzweil simply wants to say is that the brain is a computer. If we get used to this garbled language (and the concomitant garbled thinking), anything can be made to be believed. It seems that large parts of the AI community have already gotten used to it. Paradoxically that makes them so optimistic. In their view there is no limit to what computers can achieve. The use of the word "impossible" seems to be forbidden in the community.

In his community of believers, a guru cannot only get away with nonsense, he can even proclaim proven falsehoods. In the same paragraph as that of the above statement, Kurzweil declares some lines later that, "If understanding language and other phenomena through statistical analysis does not count as true understanding, then humans have no understanding either." As an example for true understanding let us take the simple question "Is it possible to fold a watermelon?", thought up by Ernest Davis (2016). The answer is easy for a human because he knows what a watermelon is and what "to fold" means. He *understands* the question. Computers don't have this knowl-

---

[25] What we are witnessing here is the knack, well-known from psychology, to throw the highly prestigious science of mathematics at problems unsuitable for its application hoping some of the prestige will rub off on the procedure.

edge but just have the words embedded in millions or billions of phrases (here the pompous term "big data" comes in), yet not connecting any meaning to them. They do not understand anything.[26]

All they can do is make a statistical analysis of the occurrence of the two words in those phrases and then make a decision about whether a watermelon can be folded or not. At this time computers are unable to answer questions like that and, considering that questions may be created that require much more complicated understanding than the above one, they most probably never will be able to answer them. In any case, it is patently false to say that humans proceed like that. They obviously need not and also cannot. They don't have billions of phrases stored, and if they had, it would take them years to browse them and then make a statistical analysis.

But with the firm belief that the mimicking of mental processes must be possible, AI research set out to compare humans and computers with respect to objective performances in highly restricted tasks. In this respect, computers catching up with humans or even surpassing them appeared possible. Computers beat human masters in both chess and Go, events touted by the AI community as historical steps toward making computers intelligent, even more intelligent than humans. But, computers surpassing humans million- or billionfold in their capacity to predict possible moves, a computer beating a master in chess appears not that historical after all. As Collins (2018) puts it: A chess program beating a human at chess is "nothing more portentous than that a tractor can beat a human at ploughing" (p. 84). The portentous thing rather is that it took programmers so long to have a human beaten by a computer in chess.

As to human-likeness, it is obvious that what is going on in the computer when playing chess is totally different from what is going on in the human mind when doing so. The computer's capacity when playing chess is largely based on its superior computing power, its "brute force."[27]

---

[26] The latest figure I came across is 200 billion words from books, articles, or websites underlying "A remarkable AI that can write like humans – but with no understanding of what it is saying" (Hutson, 2021).

[27] It has been claimed that the computer beating a Go master, which allegedly used a deep-learning program, was human-like in that it even had intuition. As it so often happens in psychology, it is hard to define intuition and we know little about how it works. But what

As mentioned, the trick of using the term Artificial Intelligence when talking of computers makes them intelligent by definition, irrespective of whether anything like human intelligence has been created. Programming computers to play chess (and beat masters) is probably the most prominent example of achievements in the field. But it at the same time demonstrates that nothing like intelligence has been created in the first place.

According to people like Kurzweil or Hawking computers will not only outsmart humans chesswise in the not too distant future, but in any other intellectual domain as well. And so according to their vision, this will inevitably lead to computers which will wonderously have acquired human motives, outmanipulating and dominating the poor humans, holding them as pets at best, if not exterminating them altogether because, stupid as they are, they have become nothing but a bore to the new intellectual giants. Kurzweil et al. even found a pompous term for this scurrilous vision: "The Singularity."[28] With their tremendous means for propaganda and a world-famous physicist on their side, AI people actually have managed to have serious people deal with that nonsense. Ideas like "The Singularity" now widely spread, both in science and the broad popular consciousness, those serious people must deal with it whatever they may think about it. And in the mental environment they now have been forced into, they prefer not to call it nonsense. As it happens they eventually even start taking outlandish claims seriously and start talking in the nomenclature presented to them by people like Kurzweil. That is where the debate about what computers can and what they cannot do has been going badly wrong. It is where, as Harry Collins puts it, humanity starts to "surrender to computers." Once computers are seen as

---

we do know is that the term may not be applied to a step-by-step learning procedure as is typical for deep-learning. A human proceeding in the deep-learning way (if that were possible) would never have an Aha-experience or exclaim "eureka" (see above). Hopefully outlandish claims such as that computers possess intuition will accelerate what, according to Collins (2018), many AI scientists fear: "[…] that yet another 'AI winter' will follow the latest deep-learning bubble."

28 The term is stolen from mathematics and cosmology in the hope that some of the prestige of those sciences may carry over with it. None of the meaning in mathematics or cosmology can be found in AI research, though.

human-like and Moore's law[29] being in effect, Singularity must, in the minds of the believers in unlimited computer power, come about by necessity.

When about two generations ago Weizenbaum conceded that computers could be socialized, based on the weak analogy that they can be modified by their experiences with their world, he made a grave mistake. He made that concession even though he thought that in any reasonable sense, i.e. not just by mere analogy, computers can never be socialized because socialization means a lot more than being affected by one's world.

Weizenbaum's critique set in *after* he had made that concession to AI speak, when he stated that the mere question in what way the socialization of a human is different from that of a machine is a sign of madness.

But the damage had been done by himself. The great computer pioneer and engaged critic of the idea of the humanness of computers had, in his own fit of the very folly he later labelled madness, conceded that computers can be socialized. His using a term he thought to be totally inappropriate in this context only because AI people did, looked as if he was prepared to argue the inarguable, to discuss something he found nonsensical. Once he had entered the discussion by conceding that machines were socializable, it became difficult to think that the whole discussion appeared nonsensical to him.

To accept and use the anthropomorphizing nomenclature of the AI community is like falling into a trap. The moment you do it, you are partaking in a discussion you actually find nonsensical. And, worse, when you partake in it for a while you may not find it nonsensical any more. Our mind somehow has a desire for meaning, for sense. The AI think being highly popular and pushed aggressively, one can hardly avoid the discussion and once you are in it the (unconscious) meaning-seeking mechanism sets in. As a result you may end up biting your nails when the latest computer program faces the latest Turing Test (see below, pp. 30) because you are not sure whether a machine can't be human after all, and the result of the test may decide about your sense of self as a human being, perhaps convincing you

---

29   Moore's law originally stated that the processing power of microprocessors doubles about every 18 months. Applied to technological processes in general it means that the increase in any technological capacity increases exponentially, not linearly, with time.

that you are nothing but a machine after all.[30] As to the intellectual capacities of computers (actually the alledgedly possible, the phantasized ones), we usually find several levels, their definition depending on the visionary capacity of the authors postulating or proposing them. The range of possible "intelligences" is huge, starting with things like the thermostat whose intelligence regulates the temperature in your room, up to "societies" of self-replicating, self-repairing, and self-enhancing machines which are like extraterrestrial aliens whose society and culture is incomprehensible to us. On the long and winding imaginary road from simple everyday machinery to the above societies of autonomous machines, a road which leads far beyond the creation of human-like ones, astonishing things happen. And because they only happen in the imagination of those who dream them up and tell us about them, those things exist but on a purely verbal basis, not bound to any rules of empirical knowledge or logic, not to speak of any relation to reality. Formulated in a linguistically correct form, stemming from assumedly serious scientists and obviously not meant to just be funny, they may often impress us as rational ideas, quite unlike Francis Picabias earlier mentioned insight about the directional change of thoughts. Seen from a strict logical perspective they often appear quite Dadaesque however, even though not as entertaining as those launched by DaDa. But expressed in the broad and highly important context of our technological future, they may rather irritate and even depress us when envisaging them to become the dominant kind of thinking about humanness. Let us take some examples from Harry Collins' critical book about the field titled *Artifictional Intelligence* (Collins, 2018)[31]. We will see that AI speak has a lot more absurdities to offer than the socialization of a machine which strained Weizenbaum's imagination nearly half a century ago.

Today, in the AI world, the socialization of a computer has not turned out to be the absurdity Weizenbaum had hoped it eventually would, but has

---

[30] In the end that need not necessarily depress you, however. Having progressed far enough in the AI community's speak and think with its admiration for computers and its sceptical view of the intellectual capacities of our fallible species, you may even feel proud of being told that you are not just a human but even a machine (see Musk's proposal to shape human thinking after that of computers, above).

[31] Note that the following quotes are not necessarily seen as reasonable by Collins, but are common notions in the discussion about the possibilities of creating human-like machines.

become a common and normal idea. Computers are not only "embedded in social context in the same way as humans" (p. 7), they can even "embed themselves within and learn from, existing human communities" (p. 58). They "will have absorbed our cultures by one method or another" (p. 79). No wonder they can learn, think and decide and many things more, all those terms used, however, as Collins sees it, "prior to any analysis of what they actually do" (p. 82). If you still doubt their humanness you are being told that they have human-like desires (p. 98) and even a religion (p. 16). You may wonder how all these functions and abilities manage to get into computers, but in AI speak that is quite simple. "Westernized human-like desires" can be "fed in" (p. 99) or a whole culture can be "poured into" artificial brains (p. 173). That feeding in and pouring into not only tells us nothing about how those things get into computers; more importantly, the phrases are nonsensical, even absurd, quite like the idea of pouring the theory of relativity into a teacup or Picabia's theory about why the head is round. However, in a community where such phrases are not seen as absurd but as quite rational, the trick of wrapping nonsense into a linguistically correct form is quite effective for propagating human-like machines. Take for example an AI critic who cannot see how a computer could possibly have human emotions, as these are strongly affected, among many other things, by the experience of a human childhood, something a computer obviously cannot have. No problem! A human childhood can easily, that is in a linguistically correct form, be fed in or poured into the computer. So, with this illusion that anything human can be fed or poured into a computer having occupied the fantasy of many AI people, we need not be surprised that they predict a human-like computer, or even find that it already exists.

As mentioned, the AI imagination does not stop with computers just having a human mind: computers may also have human-like bodies, as Hollywood has repeatedly proven, and may therefore have human-like intelligence (p. 97). With or without human-like bodies, computers will be able to replicate, to repair, and enhance themselves and thus finally be able to form "autonomous societies" completely independent from human societies. Then they will not be human-like, however, but aliens, quite like the extraterrestrial ones humanity has fantasized about for millenia, i.e. since long before there was anything like a modern computer and an AI community. As creatures lacking a basis in reality and existing in human imagination only, they are indistinguishable from all those gods and spirits dreamed up by humanity from the beginning of its existence.

Collins' imaginative capacity is not unique. Pedro Domingos (2018), a professor of computer science, cites futurists' belief that "computer models of us will become so good that they will be practically indistinguishable from the real thing," which means that we would be able to "upload ourselves to the cloud and live on forever as pieces of software, free of the pesky constraints of the physical world," which then means that "our identity might be stolen in the most absolute and complete sense of the word." He is sceptical about the possibility but typically does not want to exclude it. He is not doubtful, however, that "within a decade each one of us will probably have a 'digital double', an AI companion" (Domingos, 2018) which is "a model of you, learned from all the data you have ever generated in your interactions with the digital world." A "'master algorithm' can then be coupled with continuous capture of our sensorimotor stream via an augmented reality headset and other personal sensors and the double will grow to know you better than your best friend." At the end of the article his AI fantasy goes definitely into overdrive producing a phantasmagoria in which "everyone's doubles will keep trying to learn models of another and they will form a society of models, living at computer speeds, branching out in all directions, figuring out what we would do if we were there. Our machines will be our scouts, blazing a trail into the future of us as individuals and as a species. Where will they lead us?" I find the answer to this pompous, wise-sounding question quite simple: "Nowhere." Phantasmagorias don't lead anywhere, at least not in reality.

Is it allowed to say that a professor of computer science talks like an idiot? That depends. Once you have entered the debate about what computers may accomplish as it has evolved, the answer must be "No."

In a discussion about computers, which ostensibly have "absorbed" human culture, religion included, and are "an exact reproduction of a human in physical constitution as well as everything else" (Collins, 2018, p. 82), and into which anything human can be "fed," you find yourself in a black hole which absorbs your sense of reality, suggesting a world where anything appears possible, making people think that anything *is* possible.

Esoteric, occult, mystical ideas have always been proclaimed. They were usually restricted to small groups or "circles," ridiculed or ignored by most people and having little influence. But there are exceptions as to the number of adherents and as to influence, like homeopathy with its many believers and its billion-dollar business.³² But even homeopathy pales when compared to the human-as-a-computer belief system with its super rich enthusiasts who can spend billions of dollars and occupy thousands of people in order to do "research" in their sense. And, as to prestige, the computer-as-a-human project profits enormously from the great achievements of AI in the most diverse fields of application where it is quite irrelevant whether you consider the functions carried out by the computers to be human-like or not because all that counts is that the machine gets its job done. Typically, the computer-as-human people do not tend to make a clear distinction between human-like computers and computers just getting a job done, like cleaning your house or driving your car.³³ They rather tend to intentionally confuse the two, knowing quite well that the actual achievements, the ones that are quite helpful but in no way human-like, do somehow lend credibility to the project of creating computers that *are* like humans. The ones created, however, are able to do something humans can do in utterly restricted contexts only. The whole confusion was started, intentionally or not, with naming the field Artificial Intelligence, insinuating human-likeness of the machines created.

With man-machine analogies, weak as they may be, but omnipresent today, the idea, impression, or feeling of human-likeness of computers is deeply entrenched in the minds of today's people, the younger they are, the deeper. But a small community of people obstinately believe that there is a qualitative difference between humans and computers, which for them implies that computers can never be made human-like. They find the idea of human-likeness of computers absurd and find the discussion about this issue pointless. For them a phrase like "culture being poured into artificial brains" suffices to characterize the field as odd, esoteric, obscure. But some of them feel that they cannot evade the discussion simply because a huge majority of colleagues partakes in it, quite as if it were a normal scientific discussion. The hope of this poor minority to have the man-machine problem eventually

---

32   Homeopathic treatments never passed a double-blind trial. When such a passing had been reported it regularly turned out to have happened by mistake or through fake.
33   Let it be noted that in the actual sense, i.e. with the competence of a human, computers can not drive cars or clean houses.

solved unambiguously in their direction rests on a procedure invented by a highly ingenious mathematician who became famous for his leading role in the breaking of the Wehrmacht's ENIGMA code during World War ll, Alan Turing. Accordingly, it is named after him.

The Turing Test – what can it tell us?

The Turing Test tests whether a computer can interact with a human in a way that makes the human believe he is interacting with a human. That sounds simple, but isn't. That it isn't can be seen from the response to the computer program ELIZA created by Weizenbaum in the 1960s which mimicked the (verbal) behavior of a psychotherapist. Even though computer power at the time was minimal as compared to the one of today's computers, people conversing with ELIZA became emotionally deeply involved with the computer and, if not explicitly told that they were interacting with a machine, they felt that it was a human being they were interacting with. The program even spread the belief that it demonstrated a computer's ability to understand natural language, which remains an unsolved problem for computers still half a century later. If seen as a kind of Turing Test, Weizenbaum's procedure of having a computer respond to a patient seeking psychiatric help is a rather weak one and it was not thought at the time that the program had passed a Turing Test, i.e. had proven that a machine can indistinguishably behave human-like. It is actually quite understandable why, in the given context, the machine passed for a human interlocutor. Weizenbaum had chosen a well-known form of psychotherapy where the psychotherapist does little more than repeat the patient's utterances in a somewhat reformulated way, which made it quite easy to mimic such a psychotherapist.

But some subjects might well have suspected, due to the stereotyped and uninformative kind of answers they were getting, that they were *not* interacting with a human. A real Turing Test does not allow such ambiguities. It must allow an unambiguous decision as to whether the test was passed or not. ELIZA just demonstrated (in a particularly impressive way) our strong inclination to anthropomorphize. Let it be noted and stressed that even if a computer passes a Turing Test, the functions on which it bases its answers to the test questions (for example yes in a yes-no question) will of course not be equal to the mental functions on which a human bases his decisions. In other words, a Turing Test cannot decide about a computer being human-like or not. All it can do is show that a computer can, in an extremely restricted context, *behave* like a human.

But could there possibly be a computer capable of passing *any* Turing Test? To answer this question, we must look at how a truly demanding Turing Test can be created. It turns out to be surprisingly simple: just create tasks the solving of which requires mental functions like understanding or thinking, functions that do not run on computers. And because they don't (except in analogous form, which does not help the computer any when confronted with such tasks), these mental functions may be quite simple ones, like those you need in order to answer questions like "Is it possible to fold a watermelon?", created by Ernest Davis (2016), or "The trophy doesn't fit into the brown suitcase because its too small (large). Question: What is too small (large)?", created by Levesque (2017). The simplicity of such tasks results in a 100 % success rate for (mentally intact) humans. Such a result would have to be also achieved by computers if they were to pass the test.[34] Achievements above chance would definitely not suffice. Today's computers are close to 50 %, i.e. they cannot answer such extremely simple questions. Obviously, they are anything but intelligent. They cannot think.

Why are such tasks so simple for humans and so extremely difficult, actually impossible to solve, for computers? To answer this question, we must take a closer look at what is going on in the human and the computer when they are confronted with Davis' or Levesque's question.

We don't really know what is going on in a human in a concrete sense (brain scanners are of no help in this context), but some knowledge from linguistics and psychology may help form a simplified notion of what happens. Taking the melon task as an example, humans have formed a concept (an idea) of a watermelon, based on direct experiences with watermelons (having seen, held and manipulated one or several) and indirect ones (having seen pictures of watermelons or having read or been told about them). The concept includes all the relevant characteristics of a melon like shape, consistency, weight, etc. In short, we say that we *know* what a watermelon is. Humans cannot just form

---

[34] One may go a little below 100 % in order to account for some people being so puzzled by the stupidity of the question that they give the wrong answer.

concepts of objects but also of all kinds of manipulations that can be applied (or not applied) to them, like folding. The concept has developed from their experiences with folding all kinds of things, mostly sheets of paper, but also pieces of all kinds of foldable materal, pancakes included. Humans *know* what "to fold" means. All a human has to do in order to answer Davis' question is to combine the two concepts and, bingo, the answer is, "No, watermelons cannot be folded." The test is so simple that it takes humans just some seconds to answer it, the time depending somewhat on how puzzled they are about the oddity of the question. As we can see from the fact that modern computers are so far unable to answer the question, their approach must be a totally different one from our human one. Their problem: they cannot form concepts and combine them, in short, they cannot think. All the computer has at its disposal in the above situation are the stored words "watermelon" and "fold", but it lacks the meaning of them that they have for humans, a meaning based on their experiences with watermelons and the process of folding, experiences a computer cannot possibly have. Having stored millions or billions of phrases, all it can do is go through those phrases in search of a watermelon being folded. But neither finding nor not finding such a phrase allows it to answer the question correctly. Not finding one does not allow it to conclude that watermelons cannot be folded. And finding one does not mean it can correctly conclude that watermelons can be folded. For the computer may well find a phrase where a melon is being folded because a Dadaist (or a somewhat deranged person) may have imagined a melon being folded. Things like this have in fact happened. Salvador Dalí apodictically and famously declared: "He who cannot imagine a galloping horse on a tomato is an idiot." So a computer being asked in a Turing Test whether a horse can gallop on a tomato will, due to the great popularity of the surrealist, most probably come across a phrase where a horse gallops on a tomato. In my view, equivalents of AI speak to Dalí's statement, like the surrealistically sounding idea of culture being poured into a computer, might very well be used as Turing Tests. However, and unfortunately, there would not be agreement on the correct answer. An AI researcher might say, quite Dalí style, "He who is unable to imagine culture being poured into a computer is an idiot." The difference to Dalí? Dalí *knew* that in reality (not in surreality) horses do not gallop on tomatoes, while AI people working on human-like computers are convinced that things like culture, childhood experiences, or all kinds of mental concepts can indeed be fed into computers.

In any case, computers being unable to answer such simple (for humans) questions like the one with the melon or the one with the trophy and the suitcase, they will certainly not be able to answer ones that require more complicated mental processes[35]. Computers just cannot *think*.

But it must be asked what a Turing Test can tell us in the first place. AI peole like to think that ultimately, when the most demanding Turing Test has been passed by a computer, we will have been shown that the computer is mentally human-like. But, as mentioned, acting like a human even in the extremely restricted situation of a Turing Test in no way means that the processes that make the computer pass the test are in any way human-like. They can tell us nothing about the human mind.[36]

So, scientifically, Turing Tests make little sense. But seen as a sport the duel between testers and programmers may be quite entertaining, even thrilling. Harry Collins (2018) sees himself biting his nails "when the powerful contenders consider it time to subject their programs to the sensible Turing Tests [...]" (p. 195). But when observing the duel, let us not bite our nails for the wrong reason. The test will never decide about computers being like humans or humans being nothing more than machines. The nail-biting should be just like the one we do when watching our favorite football team on the brink of defeat. In my view the sport in this case is not quite fair, however. One team, the test designers, can dictate the use of the capacity to think, while, after more than half a century of predictions to the contrary, the other, the programmers, have, as the melon test has shown, not managed to get any of it into their machines. If they tell us that they will in the next 5, 10, or 50 years, we would be quite naive to believe them.

But there must be a warning. AI people have shown quite some ingenuity in making machines *appear* to have the capacity to think. The latest feat in misleading us about that capacity is a computer system that can take part in

---

[35] To create ever more demanding Turing Tests does not constitute "moving the goalposts," as Kurzweil complains about critics, making it impossible for the computer to win. If AI claims to create human-like computers, these must be capable of passing *any* Turing Test.

[36] Typically even people critical of AI claims do not dare to say that intelligence necessary to pass demanding Turing Tests will *never* be found in machines. So Gary Marcus (2017) finds that "human-level artificial intelligence remains very far in the future," which of course implies that it *will* eventually be created.

live debates with people (Reed, 2021). Knowing that computers cannot answer questions of the simple watermelon type we may be somewhat puzzled learning that they can "debate", which we undertand implies that they can think. Or does it? Thinking of the above AI (p. 57), based on 200 billion words that can write like humans (without understanding what it is saying), we may well imagine a computer system producing reasonable sentences in one of about 100 topics and with that partaking in a "debate." But did that system *understand* the topic and react reasonably to the arguments of the debaters? Exactly that was *not* tested in the study. The evaluation of the performance of the system was simply to ask a human audience whether they thought the system was "exemplifying a decent performance." Nearly two thirds of the audience had the impression that it was. Producing reasonable sentences about a topic may make it appear to partake in a "debate." But it tells nothing whatsoever about its understanding of the topic or of the arguments of the other debators.

Let me stress again that the decisive argument against Turing Tests is that, even if they are being passed, this tells us nothing about the machine as being human-like. The idea behind the Turing Test, that if a machine *behaves* like a human (passing the test) it *is* like a human, is reminiscent of the ideology dominating psychology for decades, behaviorism (see above). After having made most of scientific psychology sterile for about half a century it was eventually abandoned and psychologists came to the conclusion (one rational people had never abandoned) that the same behavior may result from quite different mental processes and motives.

Behaviorism in its most orthodox form can be seen at the heart of that branch of AI research which tries to mimic human mental functions. If a computer can be taught human-like behavior in the myriad of domains where humans are proficient then it must be human-like, or so the reasoning goes. Before trying to teach computers all these things, which is patently impossible anyway, one fundamental question should be answered: If a computer were really able to do all those things, yet the informational processes (termed mental processes when they occur in humans) underlying its behavior were different from the ones that occur in humans (which they would be with certainty), would we still think of it as human-like? As for most people outside AI human mental processes constitute our identity, their answer will be a clear "No." I think that the debate about the possible human-likeness of computers should have ended here. In my view to answer "Yes" is absurd, or to use Weizenbaum's term, obscene. The computer would at best behave like a human, not be one. But as so often AI enthusiasts have started a debate no reasonable person, i.e. one not inclined to debate about absurdities, would like to take part in because for him there isn't anything to be debated. Our humanness is *defined* by our mental processes. To doubt this means to question the basis of all human culture and civilization, of our concept of humanness as it has evolved over the course of human history. Nevertheless, the debate has been kept alive by endless AI propaganda. Not wanting to be part of it, however, all I can do is try to show that the project to approximate human likeness by teaching machines more and more, and ultimately all the things humans are capable of, is a hopelessly naive one. Due to their capacity to think and to transfer knowledge from one domain to another, humans can do an unfathomable number of things, with most of them lacking the clear rules that would be necessary for them to be programmed. And by far most of the domains humans are expert in are ones that have preconditions not satisfiable by machines, like living in a human social context or having a human body. To find that a machine, merely on account of being able to behave like a human in an infinitesimal segment of what humans can do (like playing chess), is therefore at least *somewhat* human-like is a purely formal and by no means sensible statement.

AI enthusiasts do not like the term "impossible," but they must be reminded that there are many impossibilities in science, like the one to build a perpetuum mobile, to make gold out of other metals or to travel faster than light. The belief that computers can accomplish *anything*, popular in the AI community, is deeply unscientific, a *belief*. It may come at tremendous costs when an unattainable goal is being set and stubbornly pursued.

Taking the brain for the human – a deprived view

In AI speak (and think), when dealing with the physical (neural) basis of mental functions, we often read phrases in which the brain, not the human, does this and that, perceives, learns, remembers, etc. The unconscious idea behind this view appears to be that, if the brain is a computer and at the same time is the sole physical basis of mental functions, then it must be possible to create human mental functions by programming computers. The brain may be the dominant substrate of mental functions (Luria called it the "organ of the mind"), but its function can only be understood through its intricate relation to the rest of the body, the neural structures outside the brain in particular. For perceptual processes there are the afferent neural fibres of the Somatic Nervous System coming from the diverse receptors of the sensory systems. For locomotion there are the efferent neurons of this system leading to the striate muscles. For sensing the internal state of the body there are the underlying afferent neural fibres of the Autonomous Nervous System, and for upholding homeostasis there are the efferent neural fibres of this system going to the effector organs like the blood vessels or, mediated by the pituitary, to the diverse glands. Not to forget that the function of the Autonomous Nervous System is intricately related to and coordinated with the Hormone Sytem. No mental or emotional processes can be understood when the functions of these systems are ignored.

The brain-body system as described does not exist in limbo. It is intricately embedded in a social-situational environment, specific for each single human individual. So no mental or emotional process can be understood when ignoring the intricate interaction of the brain-body unit with the environment. We should even speak of a brain-body-environment unit.

The intricate relation between the brain and the rest of the body need not be stressed in physiology; it is self-evident. And the decicive role of the organism-environment interaction need not be stressed in the social sciences; it is self-evident as well. But in a time dominated by the brain-as-a-computer delusion, the fact that the single human mind evolves in the process of the human (not just the brain) interacting with the environment, physical and social, is largely ignored.

Computers having just a few functios analogous (not reasonably comparable) to the ones shown by the brain and thus not being anything like the human brain, lacking a human body and not existing in anything like a human social environment, it must be asked how anyone came to think of a human-like computer. The whole idea is an absurd one and should not be debated in a scientific context.

But as it still dominates the discussion and as the absurd will be with us for a long time to come, I feel compelled to further detail why the whole debate makes no sense and to state time and again what, in times not dominated by the brain-is-a-computer delusion, would be entirely obvious.

A short phenomenology of mental functions

As mentioned, the computers-as-humans debate suffers, among other things, from ignorance about the biological, largely neural substrates of the mental functions (as we have seen) and a particularly deep ignorance about what human mental functions are in the first place. These functions are the central topic of the field of psychology, but, as mentioned too, psychology as a science, at least as taught at most departments of psychology, has contributed little to conveying a realistic picture of them. Most psychologists, intent on riding the computers-like-human wave themselves, have rather contributed to the gross oversimplifications about mental functions characterizing the debate.

Before addressing in more detail some of the functions listed under the heading of General Psychology, let us shortly remember William James' description of the mental functions in the form of the "stream of consciousness" or "stream of thought" as a constant flow of all kinds of closely interrelated and interacting mental processes (perceptions, memories, thoughts, etc.) happening against the background of diverse mental states (emotional, attentional, activational, states of consciousness, etc.). The mental processes as presented in the textbooks and addressed in the following are integral parts of that flow and cannot be studied in isolation. They should rather be seen as *aspects* from which that stream may be approached, whereby each approach to an individual function must acknowledge that it makes sense only if seen in the broader context of all the other functions happening in parallel, and, not to forget, that it should keep in mind that all those functions and states are subject to constant change, depending on the person's interior and exterior environment. A realistic description (a phenomenology) of human mental functions not only demonstrates their utter complexity but also that they are idiosyncratic, i.e. specific for every single individual at any period of time.

Memory

Having earlier embarked on an initial discussion of sensation, perception, and learning, let us now first turn to a function where the computer-human analogy most easily comes to mind: memory, the capacity to store and retrieve information, something both humans and computers have. But even if we consider just (seemingly) simple memory functions, such as storing and retrieving telephone numbers, a closer look reveals that in humans, in contrast to computers, we are dealing with a huge diversity of processes underlying the memorizing, with each and every person using different ones for doing the job. Except for the fact of information being stored, there is no parallel in humans to the stereotyped way a computer does that job. This fact has been obscured by the computer metaphor which suggests that memory processes in humans are quite machine-like.

Just a short look at how people memorize phone or any other kind of numbers immediately reveals the difference. Humans, for example, memorize numbers in different functional domains like audition (memorizing the sound of the numbers as they are being spoken), vision (memorizing them as they appear in written form), or memorizing the movement of the fingers across the keyboard when calling, which would be a combination of visual and tactile-proprioceptive functions. The numbers may be associated with numbers already memorized, like birthdays of people we know, days or years of historical events, world records in all kinds of disciplines, or sizes of all kinds of things we know, or they may be broken into units of two or more digits, etc., etc. Taking all such possible forms of memorizing and their possible combinations, the combinations used probably being different depending on what the numers stand for (telephone numbers, days of the year, hours of the day, etc.), makes for an infinite number of possible ways of memorizing numbers. As a result, not two people on earth do use exactly the same scheme. So much for *just* memorizing numbers. Obviously, for humans to use such an abundance of mental functions in order to just memorize numbers poses no problem.[37] Obviously there is a qualitative difference between human memory and that of a computer. Both may contain the same information (like

---

[37] Techniques to support or enhance memory (so-called mnemonics) have existed long before psychology started to study memory. They are not based on experimental results but

phone numbers) but the human memory, as we have seen, implies many other mental contents, too, i.e. all the associations, ideas, thoughts, feelings, etc. which give *meaning* to the numbers memorized. It is an extreme oversimplification to see a telephone number in the human mind as just that, a telephone number, i.e. a sequence of signs allowing to establish a connection between two telephone sets.

All this already may give us a glimpse at what distorting simplification we are committing in speaking of a computer's "memory" and thereby evoking the association with something totally different, the human memory.

As to be expected, things become a lot more complicated, and the difference between the human and the computer still more pronounced, when we consider the human memory contents that givepeople their unique capacity to understand and control the world, contents we already met above when elaborating on what a computer is overtaxed by when trying to answer such a simple question as to whether a watermelon can be folded; these contents are: ideas, concepts, principles, categories, thoughts, theories, etc., just everything that has to do with meaning and comprehension and is often but not always bound to language. As we have seen such things simply do not exist in computers.

Human mental efficiency, based on the formation of ideas and concepts, was quite obvious in the above watermelon example. Combining the idea of an object (watermelon) and the concept of an operation (folding) allows a human an immediate response to the question. Simple as this is for humans, a computer, not posessing anything like ideas or concepts, must go through millions of phrases searching for one that may help it answer the question posed by the Turing Test.[38]

The question is a simple one for a human, but the developmental processes that led to the formation of the concepts he uses to answer it are anything but simple. Before being able to effortlessly answer a question, even a simple or stupid one, a human must go through an arduous learning process

---

are purely pragmatic, like memorizing items to be learned by associating them with specific points of orientation when taking a walk. So when memorizing the states of Mexico, for example, the state of Morelos may be recalled via the Main Library.

[38] As we have seen in the case of ironical or provocative language (see Picabia), the computer, not having a concept for irony or provocation, must be helpless anyway.

of forming ideas and concepts, of developing routines to combine and apply them in order to understand or describe all kinds of situations, processes, functions, or events. Only through years of living in a human context, such routines develop and can be applied. The human context (parents, teachers, peers, books, media, etc.) is decisive in supplying ideas, concepts, categories, etc., developed by humankind over millennia, most but not all of them codified in language.

Watermelons and folding procedures are just an infinitely small part of those concepts, but to develop them is by no means simple. The concept of a watermelon involves several sensory systems like the visual one (which will elicit an already existing concept, that of a sphere), or the tactile system. For gauging the weight of melons, the proprioceptive system comes into play (for example the unconscious perception of the tension in the tendons involved in holding or manipulating the melon) as well as the motor sytem, necessary for holding the melon. The motor system is also necessary for finding out about the consistency of the melon (e.g. hard vs. soft) by applying pressure to the surface of the melon with one's fingers and at the same time registering the pressure at the fingertips resulting from that application, which in turn involves one of the cutaneous sensory systems, the one for pressure (the other two being the ones for temperature and pain).

In order to describe in detail what goes on in the sensory and motor systems when forming the concept of a melon, one could go on for many pages, but what has been described may suffice to give an impression of the utter complexity of those processes, not to speak of the complexity of the processes necessary for establishing the concept of folding, as it includes the concept of an object (the one to be folded) and of a function, the latter resulting from the experiences of folding all kinds of objects (sheets of paper, handkerchiefs, shirts, pancakes). Note that that utter complexity can be easily inferred from the simple fact that it takes years to form them.

When concepts are firmly established, they are routinely (and unconsciously) applied and that application appears to be quite easy. As will be seen time and again in the following, for example in the context of perception, mental processes are so typically routine that we are not even aware of them happening. This is why we so often underestimate their extreme complexity. To correctly answer the watermelon question appears quite simple to us. But even the most powerful computer, prepared by the most capable programmers, cannot give that answer because it lacks the arduous process of concept formation.

The mental representations of things like watermelons or of operations like folding are still quite simple ones when compared to the most sophisticated ones developed by humans like theories, systems, plans, schemes, ideologies, world views, religions, philosophies, etc. They cover the whole range of logical stringency, from theories in the natural sciences (most of them logically consistent) to all those -isms signifying ideologies, like Darwinism (Darwin's evolution theory falsely applied to society, an ideology), Marxism, behaviorism, positivism, existentialism, or the diverse religious systems (logically mostly quite inconsistent). They have ultimately been created in order to understand what happens in the concrete world (how planets move, or why people behave like they do, for example), but their contents are abstract ones, creations of the human imagination, not representations of things or processes in the concrete world. The conditions for them to be represented in the mind go far beyond those for the establishment of concepts like watermelons or folding operations. Before a theory, for example, can be stored in memory, a mostly long and arduous process of acquisition, of understanding it, must have happened. The same is true for all kinds of philosophies, ideologies or religions where the acquisition is not so much one of a logical structure but one of forming mental and emotional associations. It is not possible, however, to describe in any generality the process of such acquisitions because they are extremely diverse and differ between people. Even in mathematics, where you may expect an orderly mental process leading to a proof, diversity dominates. As Roger Penrose (1989) observed, for example, mathematicians working on the same problem may use quite different modes of thinking, for example a more visual (geometric) one or a more formal (algebraic) one. With their mental processes not necessarily verbalized, difficulties between mathematicians as to understanding each other may therefore arise. Obviously mathematical reasoning is anything but a mechanical, machine-like matter. When working on mathematical problems not two people do mentally the same things. Different as the mental operations leading up to a proof may be, they must comply with the principles of mathematics, however.

A lot more customary than natural scientific theories or systems are the above-mentioned -isms, systems that comprise a whole world view. Practically everybody adheres to one, because, even if mostly unknowingly, everyone has his own philosophy of life, based on his personal experiences but also influenced more or less by established philosophies like socialism, determinism, or catholicism. Common to all these sytems and philosophies in the present context (human-like machines) is that they evolved from humans being embedded in their social context, having general human and purely private experiences, and drawing general human and purely personal conclusions from them. Experiences common to all humans are, for example, the ones laid down in literary form in André Malraux' *Condition Humaine* or in Freud's more stereotyped scheme of the Ego struggling to bring the demands of the Id and the Superego in accordance with each other and with reality. This human condition, represented consciously or unconsciously in each human's mind, is by definition different from anything a computer may have stored in its "memory." A further aspect of human philosophies or world views, which makes them inimitable is their emotional content. Existentialism, for example, results from the impression that human life makes no sense, that life is *absurd*. That impression of absurdity, having originated by essentially rational observations about human life, took on a highly emotional quality, a mood which pervaded the cafés and clubs of Paris in the 1940s and 50s. Needless to say that feelings and emotions cannot possibly happen in computers because these do not just lack anything like human experiences but also the neural and humoral structures underlying emotions (see below).

Beyond the specific nature of the contents of the human memory there are some further specific aspects of the function of human memory. In the storing of mental contents in memory, interest and motivation play a crucial role. With a high interest, we learn with ease, without it, we have to labor and will memorize inefficiently. With little interest, we may still learn if we are highly motivated to learn some content just because we have to. Lacking both interest and motivation, we hardly learn (and memorize) anything.

Human memory contents partly get lost, but we don't know why. There has been a long debate in psychology, psychiatry and neurology about memory losses called "dissociations," losses of memory contents of high relevance, mostly associated with some kind of psychological trauma. According to Freud (and many psychiatrists and psychosomaticists today), mental contents may be "repressed," i.e. forced out of consciousness, as a defence mechanism against experiencing the aversive, anxiety provoking quality of those events. The explanation is a hypothetical one and will stay so. To say that some contents are *not* forgotten because they are in some way important explains little. All the above examples tell us is that complex *psychological* processes affect memory. As to the *physical* (neural) processes underlying the typical loss of memory over time it explains little to say that a kind of "decay" process takes place. This is nothing more than a description of what happens.

Mental contents change over time in unpredictable ways making the testimony of witnesses in criminal processes an unreliable matter, a fact Elizabeth Loftus (e.g. 1993) has documented in the highly sensitive context of child abuse. It may even happen that a memory content is a false one from the very beginning because the *perception* of an event has been a false one, i.e. the perceived fact has been twisted right in the perceptual process, rather than in the process of memorizing.

In humans there is a huge difference between active and passive memory, i.e. we can recognize a lot more than we can voluntarily extract from memory. The fact that we recognize a seemingly unlimited number of pictures, melodies, sounds, voices, etc. points to the huge capacity of human memory. It may even be that our total consciously experienced life is laid down in our memory (Penfield & Perot, 1963; Sukel, 2012). Differences between humans as to memory capacity are huge, with the extreme end occupied by so-called "savants," some of whom can, for example, read a book once and memorize it word by word. Nobody knows how they do it.

Let it be stressed, and the principle will repeatedly be stressed in the following, that human memories, as well as other mental functions, may not be treated in isolation. They contain information in a meaningful mental context. As Max Wertheimer, the Gestaltist who stressed the contextual view like nobody else, put it: memory contents "are not blind, arbitrary, and devoid of meaning." (Wertheimer, 1922).

AI research again and again pointing to the analogy of both humans and computers "storing information" in order to enhance the belief in human-likeness of computers, thereby ignores the fact that the information as stored in human memory and the one stored in a computer are fundamentally different. The information in human memory carries meaning, the one in the computer does not.

How can that be? Intuitively meaningless information seems to be a contradiction in itself, after all. But the very definition of information as given in Claude Shannon's (1948) *Mathematical Theory of Information* does in no way refer to the concept of meaning. It is a purely formal one. It refers to a sequence of signs or symbols, each of which carries a certain amount of information, like, for example, the signs used in written language (letters, punctuation, quotation marks, and other symbols like dash or slash plus space). If the probability of occurrence of each signal is the same (which is not the case in natural language), the information carried by one of them is equal to *ld n*, where *n* is the number of signals and *ld* is the binary logarithm, the unit for information being the *bit*. According to the theory *ld n* would also be the maximum information transmissible by the system. Thus the maximum information transmitted in a system would be a chance sequence of signs of equal probability of occurrence. For the about 32 signals used in written language there is therefore a maximum information transmitted per symbol of about *ld 32* = 5 bits. Chance sequences by definition not carrying *any* information we would here deal with the seemingly self-contradictory case of meaningless information (In natural language the information per symbol is about one bit). The above formal definition of information applies to the storage of information in computers. Other than the information in human memory it can be quantified exactly (in bit-units). The information as stored in human memory is of a totally different kind. It is information that carries *meaning*. What is it that makes information meaningful in a human being? In human memory information is stored in the form of concepts or ideas, some, but not all of which are connected to words. Take the above example of the concept of "folding." "Folding" or "to fold" has meaning for a human because it is closely connected to all kinds of experiences with the act of folding, direct and indirect ones. A human *knows* what "to fold" means, whereas the computer doesn't, as little as a dictionary containing the word does. If the computer (or the dictionary) contains an explanation about what "to fold" means, that does not mean that it (or the dictionary) knows what "to fold" means, because both the computer and the dictionary lack any experience with the act of folding. In fact, the explanation in either is put there by humans in order to convey the meaning to other humans. For *them* it is information that carries meaning. Because the information in the formal, mathematical sense, the one stored in a computer, does not contain meaning the machine is unable to answer the simple question whether a melon can be folded. The myriad of human experiences with ideas and concepts missing in computers it must be asked if the analogy between human and machine "memories," based as it is solely on the fact that both store information, makes sense in the first place. It may be of some practical use, but, on the other hand, it is quite misleading as to the nature of the computer's memory and, as a consequence, as to what a computer can accomplish.

Perception

Let us proceed with a further aspect of mental functions listed in the textbooks, one intricately involved in the storing of knowledge in memory: perception.

All human knowledge originates from perceptual processes, with the sensory systems mediating the contact with the outer world. It has been amply shown that those perceptual processes are highly active ones, largely transforming the incoming information before the actual conscious[39] perception happens (a "percept" is being formed), which is then further transformed by cognitive processes (e.g. being put in context) and eventually stored in memory. So our knowledge and all the psychological processes that connect to it are crucially affected by what happens in the perceptual process.

The most important insight from experimental work on perception, elaborated on extensively by the Gestaltists (see above), is that there are no constant relations between stimuli and the sensations or percepts they elicit, but that stimuli are always perceived in context and may have totally different perceptual effects depending on that context.[40] This is the "systems view," generally accepted in psychology today and applicable to psychological processes in general. The context may very well be a psychological one, like *expectations* about what will happen determining what will be perceived. If, for example, a player in a game is known for foul play, you will in a somewhat ambiguous situation *see* him play foul, even though he not necessarily does.

The social context may play a decisive role even in rather simple perceptual situations like, for example, in detection and dicrimination experiments

---

[39] A perception is not necessarily a conscious one. About unconscious processes see below.
[40] Best known and amply used by the Gestaltists are the optical illusions where typically an element of a figure presented has different value or meaning depending on what is being presented around it.

where the social situation (the kind of instruction given by the experimenter) may determine the subject's expectation about the probability of occurrence of the events, which then determines the proportion of hits (stating an event occurred when it actually had) and false alarms (stating that it had occurred when it actually had not).

In the context of this book (can there be human-like machines?), a subfield of perceptual psychology termed "social perception" is of particularly instructive value. The above example about the perception of foul play is taken from that field. The most influential theory that came from the field is called "attribution theory" (Heider, 1958). It is about how people make causal inferences about behavior under conditions of uncertainty, i.e. to what cause or causes they "attribute" the behavior. In a football game, for example, a dangerous collision may be attributed to an unlucky constellation of events or to a player intentionally provoking it (playing foul). As to be expected, the theory is mostly about false causal conclusions, so-called "attribution errors" and how they can be avoided.

In the real world, i.e. the one outside the social psychological laboratory, there are innumerable social situations where causal inferences are being made but attribution theory cannot present binding rules or algorithms according to which causal inferences about social behavior are made in general. Knowledge about the principle (causal inferences are made via specific cognitive processes called attributions) and about the experience with typical attributions, false ones in particular, must then suffice for the applied psychologist (a psychotherapist, for example) in his effort to help people with their false causal assumptions about, for example, the behavior of a partner.

Any human perception, even the one of a simple object, is predetermined by the perceiver's having been integrated into human society. Perceiving a chair, for example, includes the categorization of this object (piece of furniture, thing to sit on) as learned from society. The visual agnosias, neurological conditions in which the patient can describe an object but is unable to tell what it is, very clearly demonstrate this fact. Such a patient can be said to *see* an object, but he does not *perceive* it in the actual sense. With the categorization of the object lost, something exclusively developed in the human social context, the *knowledge* about the object is lost. The patient is not only unable to name the object, he also does not know what to *do* with it, how to *use* it.

The process of categorization learning (and, implicitly, the formation of the corresponding cortical representations) is anything but a logical, mechanical process. This can be seen from the fact that it is person-specific to a high degree. As mentioned before in the context of complex cells, there are general categories, like the one for faces, but every human, depending on his personal environment, also has to form group-specific ones, like the ones specific for vocational groups. And there must be purely personal ones, like the one comprising all the objects I need for doing my work or the one containing all the people I consider to be my friends. No doubt there can be no general, binding system of perceptual categories.

The idiosyncrasy (person specificity) of perceptual categories, as of all the other mental characteristics, is all but ignored in AI research. AI researchers, when contemplating the creation of human-like machines, seem to think that there are typical, kind of average, and thus simulatable, mental characteristics or functions. But each human having his own, qualitatively distinct mental development, no such typical mental characteristics can exist. Averages apply to quantitative variables, not to qualitatively different things (in the jargon of psychometrics they do not apply to objects represented on "nominal scales"). So for any human personality to come into existence, it takes the whole developmental history of a human and of a human only. As so often in this absurd debate, I must state the obvious: machines cannot have the developmental history of a human because they *are* not human. If you find this statement tautological you are quite right. It is a tautology prompted by the absurdity of the discussion.

In social perception, the principle of categorization has a strong effect as well. How we perceive a person depends, at times crucially, on all kinds of categories like political affiliation, vocational training, belief in god, being male or female, etc. The most powerful and devastating categorization is the one into racial groups, those groups assumably differing with respect to intellectual and moral faculties (Darwin's terms). This categorization is quite pervasive in western civilization, but was not always. As Snowden (1983) has shown, skin color in Ancient Rome was just that, skin color, and bore no relation to those intellectual and moral faculties. This shows (once again) that racism is a social-cultural construct and is in no way based on genetic differences between races as to mental characteristics. There are of course genetically based differences as to most *bodily* characteristics, though.

A human may perceive quite a lot of objects, but when it comes to social situations and relations, the possible number of percepts is actually limitless. Perceiving social contexts, which includes *understanding* them, is a marvellous capability that can of course only be acquired by us being brought up and living in a society of humans. Think of our marvellous ability to understand irony, for example, where we, depending on the social context, correctly understand that the opposite of what has been said is actually being meant. Perceiving a social situation is, in terms of the mental functions involved, a highly personal conglomerate of thoughts, conclusions, attitudes, expectations, memories, etc., interwoven with and affected by all kinds of feelings and emotions. It is one of the most complex kinds of streams of consciousness imaginable. Rules or principles governing social perception, like the ones proposed by attribution theory, may only exist on a highly general level. In a concrete situation, like the one a psychotherapist trying to help a client with his psychological problems finds himself in, it may be quite helpful for him to know about the mechanisms of attribution (general level), but for his actual task (helping the client) there are no known rules, comparable to the ones an engineer has at his disposal, that would allow him to understand the client's problems and enable him to treat him successfully. For that he has to rely on his experience as a therapist, common sense, intuition, empathy, etc., i.e. all that is necessary for understanding the client's problems in his human context but is absent in computers.

The absence of binding rules for social perception (like for most other mental functions) may only surprise people with a naive and simplistic view of mental processes in general.

Someone intent on creating human mental functions in computers will necessarily take such a simplistic view. If he started with James' phenomenology (think of all those cognitive, motivational and emotional processes playing a role in human memory function for example) he would probably give up his intent altogether (how, for example, emulate the emotional aspects?). All he can do is take some computer functions which somehow resemble or are analogous to human mental functions, hoping that over the years an approximation to the human mental functions will be possible. Such computer functions are quite simple ones, simplistic compared to what is intended, like the algorithm helping to find a way out of a maze described above. When over the years the approximation does not succeed (so far not one has ever succeeded and it never will)⁴¹ in order to not concede failure he will *declare* that the functions created are human-like. Which is what we usually find when reading about the latest claims propagated by AI research.

The status of the flawed idea of human-like computers has even shifted from shaky hypothesis to axiom. But this shift was not due to logic or to empirical evidence as we would expect from a scientific development, but was pushed against any scientific reason by a gigantic industry and thousands of highly motivated, computer-happy young people and supported by the most professional of all creators of illusions, the Hollywood filmmakers. No journalist and hardly any scientist can withstand this pressure and state the obvious: there *are* no mental processes in computers.

Let it be stressed that the purpose of this book is not to push the status of the idea of human-like mental processes existing in computers back to one of shaky hypothesis. I want to make clear that the idea is not shaky but that it is plain nonsense. This is an important distinction. In science a hypothesis, shaky as it may be, is an assumption that can eventually be verified (or falsified). To ascribe the status of a scientific hypothesis to the idea that there can be human mental processes in machines would mean that it is scientifically debatable. But to debate it means to produce further nonsense, quite like the debate Weizenbaum maneuvered himself into when he accepted the phrase that a machine can be socialized.

Motivation and emotion

Among the psychological traits, states, and functions listed in General Psychology, motivation and emotion⁴² are specific in that they are intimately connected to neural and humoral structures we usually do not link with the cognitive functions like memory, perception, thinking, or language. Those structures are the autonomous nervous and the hormone system which regulate changes, for example, in the cardiovascular or the digestive system,

---

41   As Bertrand Russell (1959) put it, "Each of us, if he is not a dogmatic behaviorist, believes that in our innermost there are processes that can *never* happen in a machine." Note that Russell has no problem with using the term "never."

42   In the following I only use the term emotion. Emotions may often act as motivational entities, like pain which motivates us to avoid lesions or to seek therapy.

changes that bear a close relation to emotions, which is why they play an important role in theories of emotion (even the decisive role in the oldest theory of emotions, the one independently proposed by William James and Carl Lange). The hypothalamus is the central, both neural and humoral structure regulating the function of the cardiovascular and digestive systems. Well known, even to a wider public, is the role of the hypothalamus-pituitary-adrenal axis in the stress response, called the "fight-flight" response by Cannon, who first described it. As always the close interrelation between psychological functions must be stressed here. If, for example, we consciously perceive (which is a cognitive process) the increased cardiovascular activity in a stressful situation (an increased heart rate and a stronger heartbeat), that may enhance our feeling of being stressed and, depending on the intensity of that feeling, may even drive us into a panic attack. Or, when, upon feeling a specific pain, we suddenly develop the idea that the cause may be a serious one (like cancer), the quality of the pain sensation may change for the worse. The effect of thoughts (cognitions) on emotions has systematically been studied in psychology, which has resulted in the "cognitive theory of emotion" and eventually led to several techniques in psychotherapy, one of the most prominent today being the "Rational Emotive Behavior Therapy" (REBT). Its fundamental idea is that our emotions do not primarily depend on the objective circumstances we are in, but on the way we perceive and interpret them. The technique has been expanded to pain therapy (e.g. Turk, 1990) where the patients may learn to change their attitude toward the pain felt, thus, for example, becoming more confident of being able to cope with it. Pain patently exemplifies the interaction between different psychological functions in creating an emotion, functions based on neural and humoral processes in different parts of the brain as well as ones outside the brain. Modern pain theory distinguishes between a sensory, an emotional-affective, and a cognitive-evaluative aspect of pain. The affective aspect is the one that gives pain its unpleasant character, of anxiety, for example, in acute pain or of depression in chronic pain. The neural structures underlying this aspect are subcortical ones like, for example, parts of the limbic system. The cognitive-evaluative aspect has the neocortex as its neural substrate. The sensory functions happen outside the brain, starting in the periphery with the damaged or strained tissue from where two differnt systems of neural fibres lead to the spinal cord where they combine in a way that determines how much neural activity is projected to the brain, which ultimately determines how much pain is being felt. The neural network underlying that combination and thus codetermining the amount of pain felt may for its part be influenced by the cognitive-evaluative functions happening in the cortex. Let it be noted that the actual, concrete neural processes underlying pain are still largely unknown. Even the above-mentioned spinal cord mechanism (called "gating mechanism" because, like a gate, it regulates how much neural activity is allowed to proceed to the brain), postulated in 1965 (Melzack & Wall, 1965) is still hypothetical. The decisive role psychological processes may play in pain experience is documented by paradoxical phenomena, like the absence of pain after serious injuries immediately following accidents or in combat, or, most paradoxical of all, the childbirth pain felt by the *fathers* in certain ethnic groups (Kroeber, 1948).

Altogether there are always the most intricate interactions between emotions and cognitions which may create unique effects exerted on the stream of consciousness, modulated by the social and cultural context and, not to forget, by the personal experiences beginning at birth or even before.

The stream of consciousness with its permanent interactions of all kinds of cognitions and emotions is a subjective process which can in principle not be objectified. Parts of it can of course be communicated like, for example, in the psychotherapeutic situation where success depends on the therapist understanding the client's problems. This communication may at times be easy, like when we tell others about common, well-known everyday experiences, but can also be difficult or near impossible, particularly if people from different social or cultural backgrounds meet. But even within a socially and culturally homogeneous group there are vast differences between people as to their emotional response to identical social situations. In psychological stress research, for example, it has been tried to quantify the stressfulness of a situation a person is in by objective events (losing ones job, moving to another city, etc.), but every psychologist working in clinical psychology or psychosomatics knows that this is of little help, because each patient has developed his specific coping techniques during the course of his life, which means that the same situation may be felt as quite different by different people. The most intricate (and often daunting) problem the psychologist in an applied field may be confronted with is the fact that a report about a subjective state may be intentionally or unconsciously biased. Physicians and psychologists working in a pain clinic have ample knowledge about the relation between bodily states, mostly diseases, and experienced pain, but that knowledge is by no means complete. In low back pain, for example, there is in the majority of cases no medical diagnosis, such that it may be up to the poor psychologist to decide whether the patient actually is in pain or, for example, just wants to go into early retirement. "Poor" psychologist because there is no psychodiagnostic tool available for detecting lying, most certainly not for unconsciously biased reports.[43]

How does what has been said about emotions and pain relate to the idea of computers emulating psychological functions? No doubt emotions are an integral part of the human psyche. And they are particularly well suited for demonstrating the absurdity of the machines-like-humans project. Note that I am once again not saying that they pose particular difficulties for the project. Difficulties may eventually be overcome. An absurd project cannot succeed on a-priori grounds.

Let me address just two aspects of human emotions as furthr proof of this point. And let me stress that the project to mimic human memory, perception, or thinking by a computer is no less absurd than the one to mimic human emotions. But the absurdity may be felt more readily by the reader trying to imagine emotions in machines. For, while it is quite common to think of machines having a memory, for example, it is not that common to think of them as having emotions. While an analogy, and only an analogy, can be seen between what is happening in the brain and in a computer with respect to cognitions, it is difficult to see how human emotions could ever happen in a computer. Cognitions and emotions, thinking and feeling, are intricately related, yet qualitatively different psychological constructs. I need not stress the (often dominating) relevance of the emotions for human life. So for computers to be human-like, they would have to show the ultimate absurdity – of having emotions.

As so often in this book much of the following will be stating the obvious, usually introduced by "needless to say". But what can I do? As long as matters of fact, even the most self-evident ones, are ignored or even denied, they must be repeated and insisted upon. Put in context in the appropriate way, they reveal the absurdity of many claims having now been made and

---

[43] The techniques developed in "lie detection," today termed "detection of deception," are too unreliable for basing highly consequential decisions on them.

believed for many decades and thus become part of today's common (false) knowledge.

In the brain-as-a-computer metaphor what is meant by "brain" actually is the neocortex because cognitions happen essentially there. They are affected by subcortical structures, but not based on them. As we have seen the cortex plays a role in emotions, too, but the largest part of the biological substratum of emotions is to be found in subcortical structures and structures outside the brain, both neural and humoral ones. Among them are the above-mentioned structures active in the response to stress which reach out to the so-called effectors, like the blood vessels or the bronchi. From those organs, sensory fibers go back to the brain so that we can sense the changes in these organs, a perception which in itself is part of the emotional experience.

Needless to say there is not anything like the neural and humoral structures underlying emotions in a computer, not even by any stretch of analogy. Emotions being (needless to say) an essential and integral part of human life, this makes the oddity of the machines-like-humans idea all the more obvious.

Let it be stressed again that with respect to emotions this oddity is just more obvious. The idea to emulate human *cognitive* processes is equally absurd, but this absurdity has been obscured because the idea has been propagated by influential people for so long that it has been accepted by many as quite a reasonable one. Those propagating it have, for good reasons, avoided painting a realistic picture, both, of the nature of psychological functions, and of the differences between the neural structures underlying cognition in humans and the physical-electrical basis underlying the information processing functions in a computer. Instead they have painted human cognitions after computer functions. For these they have used terms that had been created in order to describe human cognitions (like memory) thereby making themselves and others believe that human cognitions could happen in computers. As to the analogy they drew with respect to the working elements in computers and those working in the brain, all we can say is that they are elements. But the elements in the brain, the neurons, cannot reasonably be compared to the ones in computers, the function of which can exhaustively be described as being in one of two electrical states. Neurons with their thousands of synapses may be in all kinds of electrical/chemical states which must be described in analogous form, like the state of depolarization at the axon hill where action potentials occur (and run along the axon) if a certain state of depolarization is effected by the summation of *graded* potentials. Needless to say that the analogy often drawn between computer elements and neurons, based on the fact that the former may be in one of two electrical states and that at the latter an action potential may or may not happen (all-or-nothing principle)[44] is an extreme oversimplification ignoring the other countless possible processes that occur in, on, and around neurons, most of which are still unknown today.

Learning

A further human-computer analogy beyond computers having a "memory" and being able to "perceive" things is that they can "learn," "deep learning" being the latest AI bubble.

As always, I will not start with describing the latest computer accomplishments and thereafter discuss how close they have come to ones performed by humans. Instead I am going to describe, as far as that is possible with the existing knowledge, what the psychological function to be emulated, learning, actually is in the first place and from there proceed to judge the meaningfulness of the emulation project. A debate about the closeness of the computer accomplishments in relation to the human ones is an endless one, the endlessness caused by the AI people creating ever new accomplishments allegedly still closer to the human ones with that ever-closer implying that the goal, the human-likeness, can eventually be attained. But this is flawed logic. It is, for example, not reasonable to assume that a human will eventually be able to jump five meters high just because the world record in the high jump has been steadily increasing. This comparison is not extreme. The above-mentioned robots (allegedly) playing football are as far away from playing like Ronaldo as any track athlete is from jumping five meters high.

During the decades psychology was dominated by behaviorism, from the 1920s to the 1950s, learning in psychology was not what one might expect it to be. It was "conditioning," happening in the form of "paradigms," classical (Pavlovian) and operant (Skinnerian) conditioning. Nearly everybody has

---

[44] To be exact, the processes that happen at the membrane of the axon when an action potential is elicited must be described in analogous form, too.

heard of Pavlov's dogs salivating to the sound of a bell after that sound had several times been presented together with food. Skinner's doves or rats learned to press a lever or to do more complicated things like pushing a table tennis ball toward each other (they were supposed to play table tennis), the behavior having been shaped by reinforcing minimal behavioral steps. Conditioning actually is a form of learning, but what is being learned is not mental contents like learning a language, math, or geography, but bodily functions. Pavlov's dogs learned to salivate to a stimulus they did not salivate to before, and Skinner's animals learned behaviors they had not shown before. In general terms, classical conditioning relates to the function of the smooth muscles and the glands, and operant conditioning relates to the function of the striate muscles. The former functions are mediated by the autonomous nervous and the hormone system, the latter ones by the somatic nervous system. As mentioned before, the former serve to uphold homeostasis, the latter ones make the individual perform all kinds of body movements, gesturing and miming included. No doubt both kinds of bodily functions deeply affect human mental processes. As mentioned, the function of the autonomous nervous and the hormone system, for example, intricately relates to emotions. The movement of the body, the head, and the eyes, for example, is important for visual perception, the movement of the facial muscles in miming is an important factor in communication.

Needless to say that all the functions described in the context of conditioning and their effects on mental processes cannot happen in computers. Computers do not possess an autonomous nervous or a hormone system, no somatic nervous system and thus no smooth muscles, no hormones, and no striate muscles.

Learning in the more common sense, i.e. the learning of mental contents and capacities, comprises all kinds of mental functions, thinking included, and thus does not allow any binding definition which is not trivial. How diffuse the meaning of "thinking" is can be gleaned from James' speaking of the "stream of thought" (as an alternative to the stream of consciousness), which implies that all mental processes are included, or from Descartes' "cogito ergo sum" (I think and therefore I exist) where "cogito" may also just mean being conscious. In this sense Descartes is quite modern, de facto saying that the identity of a human is defined in terms of his mental processes, an idea which has led to the definition of death as brain death in modern medicine. Without a living brain, so the modern thinking goes, there are no psycholgical processes going on any more, the processes that constitute a person's identity.

All that said, the term "computer learning" can express but a faint analogy and should therefore be avoided.[45] This is also true with respect to the "latest deep learning bubble" (Collins, 1018, p.26). Collins uses the terms "pattern-matching" or "precedent-based processes" instead of deep-learning because he does not want to anthropomorphize the process underlying the algorithm (p. 11). There may actually be, as happens in deep learning[46], step-by-step learning processes in humans, even though we don't really know if they exist at all. But such processes should not be seen as processes comparable to the ones typically occurring in human learning which encompasses the totality of human mental processes and is characterized by meaning, understanding, and insight, things alien to computers as is readily demonstratable with downright primitive Turing Test questions.

Consciousness

Descarte's "cogito" and James' "stream of consciousness" lead us directly to an ever-popular debate now having gone on for many decades, the one about the definition of consciousness, and, unavoidable in the world of AI absurdities, about whether machines can be in that state. After what I have said so far, you will probably (and correctly) expect that I deem the latter part of the debate nonsensical, even though such luminaries as Francis Crick, co-discoverer of the structure of the DNA, participated in it. The topic is highly popular among all kinds of scientists, particularly among ones from the neurosciences, but also among some coming from other fields, such as philosophy, who feel obliged to join the debate in order to help solve that fundamental problem of human existence. The popularity of the field is additionally furthered by the fact that, beyond its epistemological fascination, the concept of consciousness

---

[45] I am realistic enough not to expect that to happen in the face of AI speak dominating modern language.
[46] A profound critique of deep learning as a means to emulate human learning is given by Marcus & Davis (2019). As to concrete problems with the deep-learning algorithm see Heaven (2019).

is of utter practical relevance, for example in jurisdiction where assessing the responsibility for an act is intimately related to that concept.

As with the other mental functions before, I will not join the debate but start with the phenomenology of what we are talking about. From that phenomenology the absurdity of conscious computers will leap to the eye all by itself.

In science one usually starts with a definition of what is being studied, according to the simple rule that you should know what you are studying. We will see, however, that such a definition is still absent after more than 150 years of intense research and thinking about consciousness. However, as mentioned, the concept of consciousness is, beyond philosophical fascination, of utter practical importance and thus must be dealt with irrespective of theoretical-definitional problems connected with it. In jurisdiction, for example, decisions must be made, most of them not on the basis of clear definitions and rules established by science, but on the basis of such shaky things as reason, common sense, plausibility or intuition, where the only discernible rule is the one to take all aspects of the individual case of an act into consideration and weigh them in the context. Law people may profitably participate in conferences about consciousness, but they profit less there from the latest neuropsychological or evolutionary psychological speculations and a lot more from discussing with colleagues who have the same practical problems (deciding about responsibility or guilt). As to the psychological speculations, they may even be advised to *resist* taking them into account. When deciding about penalties in murder trials, the law community did, for example, for good reasons successfully resist taking into account the alleged scientific fact that there is an evolutionarily developed and thus hereditary basis of the tendency of men to kill men (Daly & Wilson, 1982). This tendency would be an unconscious one, and should therefore, as evolutionary psychologists have argued, be seen as a mitigating circumstance for a man who has killed a man.[47]

So what is consciousness? Psychology today typically going after the biological substrate of mental states and functions, we must be careful not to fall into the two consecutive traps we learned about above, reification and local-

---

[47] It is to be hoped that the law people, confronted with the latest "scientific" results to be taken into account, will keep their sceptical attitude toward those results which so far have always turned out to be speculations rather than facts.

ization, i.e. seeing consciousness as a thing[48] and then asking from our experience with things where it is.[49] But consciousness is not a thing but a construct, a particularly vague one at that, and, being an abstraction, there is no locus where it can be found. There are at best neural processes affecting a state of consciousness that may be located in the brain. Ever since Franz Joseph Gall presented a map of the brain showing where all the mental traits that have been hypothesized can be found, psychologists and brain scientists have fallen into the reification/localization trap, probably because it is easier to imagine a concrete thing than an abstract concept or hypothetical construct. Reification has become a kind of unconscious tradition, most disastrously in the context of the most prominent of psychology's constructs, intelligence, where it led to a biologistic and even racist view of that construct (see e.g. Velden, 2014). In consciousness research, the tradition is well alive, some of the most prominent consciousness theoreticians leading the way into this dead end. So Christof Koch asks, "Do some special 'consciousness neurons' have to be activated?" or, "In which brain regions would these cells be located?" (Koch, 2018). Twenty-five years earlier he had predicted that within 25 years (i.e. in 2018) "someone would discover a specific signature of consciousness in the brain" which he thought would be a small set of neurons characterized by a small list of intrinsic properties, "less than 10 of them" (Snaprud, 2018). His latest guess about the neural correlates of consciousness is "a hot-zone in the posterior cortex," an idea in contrast to one of the most popular consciousness theories (see below). The idea he, together with Francis Crick, had favoured for a while, that a sheet-like structure beneath the cortex called the claustrum is crucial for consciousness, was abandoned, however. Crick later postulated "thalamo-cortical loops" underlying consciousness, more recently being joined by Anil Seth (Blackmore, 2018). With no consensus in sight, we are left with the insight that consciousness must happen somewhere in the brain.[50] But in what *kind* of brain it must be asked,

---

[48] "What is consciousness made of?" *New Sientist* typically asked on its front page of 23 June 2018.
[49] As we have seen, William Uttal has written a whole book about the futility of trying to localize cognitive processes or states (Uttal, 2001).
[50] For an overview of the diverse theories see Blackmore, 2018; Blackmore and Troscianco, 2018.

because, according to some researchers, the emergence of consciousness dates back hundreds of millions of years to a time when the existing brains were quite different from that of homo sapiens. Todd Feinberg and Jon Mallatt see consciousness arise some 560 to 520 million years ago with "nested and nonnested" neural architectures (Blackmore, 2018). Bernhard Bars, author of one of the currently popular consciousness theories, sees the emergence of consciousness coming with the mammalian brain, i.e. some 200 million years ago. Steven Mithen sees it coming with an assumed "cultural explosion" some 60 thousand years ago. And finally Julian Jaynes could not find any words for consciousness in the Ancient Greek epic *Iliad* and concluded that there was no consciousness before 3000 years ago (Blackmore, 2018).

While we must ask why homo sapiens from his very beginning some 300 thousand years ago should *not* have been conscious (during all those millenia he had a body and a brain practically identical to ours), the question about consciousness in species way back on the evolutionary path has always been asked, mostly by owners of domesticated animals. As a scientist, the philosopher Thomas Nagel made the point by asking "What is it like to be a bat?" (Blackmore, 2018). The question reminds me of one typically asked of color-blind people (or ones with a color deficiency): "But how do you see the world? Does it all look grey to you?" The question comes up nearly automatically (I once had it even asked by an ophtalmologist), even though it does not make any sense. Someone who has never seen colors cannot know what grey means, quite as someone with intact color vision cannot tell a color-blind person what he sees. He is *principally* incapable of telling what his color sensations are.

So Nagel's question makes no sense either. Even if a bat had some kind of consciousness (whatever that is, see below), that consciousness would largely be determined by characteristics which are absent in humans, like being able to fly or to orient in space by a radar-like system. So a human cannot *possibly* know what it is like to be a bat. Or a mouse, a cat, a chimp or whatever animal. I think to ask an unanswerable question like the above one shows our uneasiness with the concept of subjectivity. We rightfully assume, one may even say we know, that there are sensations, feelings, thoughts, etc. in others, animals included, and we somehow want to know what these are. This is where the uneasiness arises because we feel that the access to the experiences of others is precluded, precluded on a-priori grounds. As members of the same species we of course have some indirect access to the experiences of others, assuming that in a given situation others have experiences identical or at least similar to the ones we have in the same situation. That assumption works fine most of the time, but as we very well know it can be totally wrong. Humans, much more so than animals, differ vastly with respect to each individual's mental background and thus with respect to their conscious experiences. But, sense or nonsense, humans will continue to muse what their dog is feeling at the moment or what the cat had in mind when it held a front paw into the air and looked at it for a while.

When it comes to definitions and theories of consciousness, things do not look better than with the ideas about its localization in the brain or its emergence in the course of evolution. Theory building in consciousness research is a procedure quite of its own. It does not start with trying to put generally accepted facts into a logical frame, a theory. The different consciousness theories are about totally different, mostly hypothetical things. The only thing they have in common is that they claim to explain consciousness. The most popular ones (at this time) are the "global workspace theory" and the "integrated information theory." According to the first, proposed by Bernhard Baars (Blackmore, 2018, p. 44), human brains are structured around a hypothetical "workspace" (kind of a stage in the "theater of the mind"). Mental contents that make it into this workspace are broadcast to the rest of the (still) unconscious brain. This broadcast creates consciousness in the individual in which it happens. The second theory assumes that in a "system" information is both differentiated into parts and at the same time unified into a whole. The creator of the theory, Giulio Tononi, claims to be able to determine an index representing the amount of consciousness in a system, based on the latter's complexity. Consciousness, when seen as a quantitative trait, can be found, even though little of it, in simple creatures and, differentiation and wholeness being also attributable to machines, in computers. This is not possible according to the global workspace theory which postulates that a highly developed brain, one with a global workspace "architecture", is necessary for consciousness to emerge.

A further theory, one starting from other, totally different assumptions, is called "multiple drafts theory." According to this theory, the brain is seen as a system of parallel processes (perceptions, thoughts, etc.) with *no* central theater. These processes are unconscious. Consciousness emerges when the processes are combined and lead to a response. The creator of the theory, Daniel Dennet (Dennet, 2017) is also the creator of the "meme", a unit in analogy to the gene, but applicable to *cultural* evolution. Like genes, memes can be more or less successful with respect to being carried on in the cultural evolutionary process. They are *mental* units, such as ideas, inventions or fashions. An unlimited number of them may be imagined and they cover the whole range of possible human creations from the wheel to Charleston music to the German social insurance system. No doubt there can be no binding rules about how they work in cultural evolution[51], memes being, according to Dennet, the units of analysis in cultural evolution. "Human consciousness is *itself* a huge complex of memes" (Blackmore, 2018, p. 45). According to Michael Graziano's "attention schema theory" a brain can describe itself as having conscious experiences when it links a hypothetical construction of a simplified model of how and to what it is paying attention, to a model of self.

A further idea about consciousness is the one proposed by Patricia and Paul Churchland (Blackmore, 2018, p. 45) that consciousness is not *based* on the firing of neurons but that it *is* the firing of neurons.

There are still many more hypothetical ideas about what constitutes consciousness, creating an unsatisfactory, even frustrating situation where some theories bear some similarity to each other, some are quite unrelated and some contradict each other. In a situation like this the desire may come up for a radical solution, a kind of sword to cut the Gordian Knot of theorizing about consciousness. And as is not rare in such situations, the sword is supposedly being found in reanimating an old idea, one popular long ago and then forgotten. Thus we learn from a recent article in *New Scientist* with the title "Here. There. Everywhere?" that according to a mathematically updated version[52] of the Integrated Information Theory we may find consciousness in everything from elementary particles, to crocodiles, to the universe, the theory being fully compatible with panpsychism, the old mystic notion that there is a soul in all matter, dead or alive (Brooks, 2020).

---

[51] About genes, to which the memes are just functionally analogous, we know, for example, that they participate, by coding for proteins, in the creation of bodily traits, which are more or less advantageous in natural selection.

[52] Again note mathematics being used as a prestige booster.

I think the idea might put an end to the endless debate about the nature of consciousness. Not in the way that it tells us anything about consciousness, but by implying the unwitting confession that the debate is not, and cannot be, a scientific one in the first place. We are not dealing with scientific theories here, i.e. with verifiable or falsifiable concepts. In my view we are just dealing with speculations, rather wild ones at that. From here to testable hypotheses, or even theories in a scientific sense, is a long way to go. Or is there even a way at all? Can there be a way to something undefinable? (See the phenomenology of consciousness below).

Put in perspective, we have, after decades of intensive research and the use of expensive heavy machinery and much computer power, and after thousands of articles and dozens if not hundreds of conferences, a situation in consciousness research which must, even under relaxed standards, be called chaotic. The only discernible consensus among scientists in the field seems to be that there should be more research and a lot more consciousness conferences with their lively discussions after the day's program has ended. But what if the extremely unsatisfactory situation in consciousness research, where, after so much time, money and brain power has been invested, anyone may nonetheless put together his own (untestable) theory, what if this situation is not due to too little research and too little lively discussion, but to the fact that the stated question (what is consciousness?) is of the unanswerable kind? We are lacking anything like an agreed upon definition of what we are talking about which is binding enough to be used in a scientific process. If everybody has his own definition (or none at all), then no wonder he comes up with his own theory of consciousness.

With a definition absent and most probably even impossible, yet not wanting to give up a scientific study of consciousness, all we can do is to resort to the approach William James took to psychological functions in general and elaborate a phenomenology of consciousness, i.e. a description of all the phenomena commonly connected to it.

Let me stress that such a phenomenology must carefully avoid any human-computer analogies. The debate about human-like states or functions in computers has suffered all too much from lightly drawing such analogies in advance and not first getting a realistic impression of the daunting complexity of what these states and functions are. If you start with the firm conviction that human mental functions can be mimicked by computers you will inevitably end up with *declaring* some computer functions to be human-like. This thinking is circular, of course, and will not get us anywhere.

Consciousness is both one of the most important and at the same time one of the most elusive of psychological constructs. It is important because it is indispensable as a concept in all cases where liability, responsibility or guilt must be assessed, both in jurisdiction and in everyday life. In combination with that, its elusiveness often makes for long and heated debates and controversies.

Consciousness cannot be thought without the unconscious, a term with two distinctly different meanings. The first denotes a state in which mental processes are absent, as in deep coma. This state may be seen as the endpoint of a scale reaching from a state of high alertness to ones of lesser and lesser alertness like tiredness or the states experienced when falling asleep, down to the comatose one. This scale may not be seen as a one-dimensional, continuous one, however, because there are *qualitative* differences between states on different levels of consiousness, like between the state of high activation and the one of relaxed attention, which are reflected in qualitatively different EEG patterns. As a consequence, there are no valid means for determining the degree of consciousness in a given situation.

In neurology the problem is most apparent (and serious) when brain lesions have to be diagnosed as to their effect on consciousness, and where no general consensus about how to do this exists. The *Glasgow Coma Scale*, a neurological scale for measuring a person's consciousness state used worldwide, based on three indicators of awareness (opening of eyes, speech output, and motor responses) need not lead to an unequivocal diagnosis. It is, for example, often unclear exactly what the often-used term "minimally conscious" means. The problem boils down to the fact that it is hard to get neurologists or psychologists to agree on what consciousness is (see Owen, 2017).

The second meaning of the term "unconscious," the one made so popular in psychology by Sigmund Freud, may be even more elusive than the conscious mind, but no one denies that it exists. Even the hard-nosed physicist and physiologist von Helmholtz used "unconscious inferences" (his term) in order to explain some perceptual phenomena, like, for example, brightness or size constancy. Great differences exist as to the relevance attributed to unconscious processes for mental life in general, however. In any case, the phenomenology of consciousness must include the unconscious in order to get a realistic picture of the phenomenon. And it must, of course, include the myriad of possible interactions between conscious and unconscious mental processes. A detailed discussion of unconscious traits and functions and their relation to consciousness will be given below.

The problems with consciousness as to its definition and function become immediately evident if we look at the huge diversity of states commonly termed "states of consciousness," including the shadowy world of "altered states of consciousness" where it remains unclear whether what is going on in the mind is conscious, unconscious or something in between.[53] That diversity, combined with the absence of anything like a valid theory of consciousness, makes it definitely clear that there cannot possibly be a scientifically useful definition of consciousness. A definition we might think of would either be too broad to be scientifically useful, or too narrow to make reasonable use of the term in real life.

Before going into a detailed yet still largely incomplete description of states of consciousness, let me, once again, stress the dissatisfying fact that states of consciousness, like all mental states, are so-called "constructs", i.e. entities not directly accessible. In psychology, as well as in psychiatry or neurology, the problem is circumvented by the earler mentioned procedure called "operationalization," meaning that an objective measure, one which is thought to be highly correlated with the construct, is taken to represent it, like a test score taken in psychology to represent, for example, scholastic aptitude or, as in the above mentioned Glasgow Coma Scale where a score based on opening of eyes, speech output and motor responses is taken to represent a person's state of consciousness. The latter procedure is particularly problematic, even though at times unavoidable, when one tries to assess consciousness because, as mentioned, there is not a single scale ranging from "unconscious" to "highly conscious" but a huge number of qualitatively different forms of consciousness instead. When in neurology the effect of a brain lesion on a patient's state of consciousness is to be diagnosed, a quantitative approach is unavoidable, however, because for the further treatment of the patient it is important to know *how* conscious he is. In the present context of a general phenomenology of consciousness, the quantitative approach is of little help, however, because qualitative differences dominate the picture.

---

53   I use the term "in between" for the sake of simplicity, actually meaning all kinds of interactions between, and combinations of, conscious and unconscious states.

The first thing we notice when trying to describe consciousness is the overwhelming variety of states connected to the term even when we just consider the normal ones, i.e. the ones not caused by pathology or the use of drugs. This is so because consciousness is affected by many factors resulting in differences between states both in degree and quality. Our state of consciousness is a different one when we are highly activated as opposed to being in a relaxed attitudinal state. Depending on the activational state, we will react differently to identical situations. Falling asleep creates qualitatively different states of consciousness. Going through intermediate states between wakefulness and sleep we may not be sure whether we are awake or dreaming. Dreaming is a state of consciousness qualitatively different from wakeful consciousness. While dreaming we are conscious (if woken up we can tell the content of our dream), but we are in little if any control of what is going on in our mind. The contents of our dreams appear somehow like created outside of ourselves.

It is well known that powerful emotions can affect our state of consciousness, making us respond differently to the same situation depending on the quality of the emotion like fear, anger, joy, or a depressed state, and its intensity. The nature of hypnosis and the hypnotic state is still controversial more than 150 years after Charcot, Breuer, or Freud, but we are obviously dealing with a specific state of consciousness. A person in hypnotic trance, when told that the water feels comfortably fresh will keep his forearm in zero-degree water, which will normally feel intolerably painful after about half a minute, without consciously feeling pain. Pain research is full of reports about no pain being felt in situations normally experienced as extremely painful (Melzack, 1975). They all appear to deal with trance-like states of consciousness, the ones we know from "western" science about hypnosis or from states induced by meditation as we find them in "eastern" religions or philosophies.

Nearly all of us have experienced the powerful effect on consciousness rhythmic music may exert, some have even experienced the trance-like state that can result from rhythm alone, as from the sound of drums[54] Catholics may experience a trance-like state when taking part in a liturgic procedure

---

54   Even the feeling, mentality, consciousness of a whole people may be characterized by it. In his wonderful *Island in the Sun* Harry Belafonte sings: "I hope the day will never come when I can't awake to the sound of drums."

where the congregation responds in endless repetition to the precentor's words.

All religions have created their own rites for inducing a religious feeling or consciousness which transcends the earthly, material world. They are all fine examples for the limitlessness of human ingenuity when it comes to manipulating our state of consciousness.

From hypnotic and trance-like states there is only a small step to the most intriguing problem we face when we aspire to arrive at a phenomenology of consciousness: the unconscious. The French psychiatrists and neurologists of the second half of the 19th century tried to use hypnosis as a means of retrieving "dissociated" mental contents in patients with symptoms of "hysteria." As they saw it, mental contents like, for example, the memory of a traumatizing event, can become unconscious, i.e. dissociated from the normal flow of consciousness. Freud, who went to France in order to study hysteria at Charcot's clinic, further theorized about hysteria, a disease characterized by altered voluntary motor or sensory functions, like being unable to walk or being blind, without any neurological diagnosis explaining the symptoms. As we have seen in the context of memory losses, Freud assumed that the traumatic mental content is being "repressed" into the unconscious due to its anxiety provoking character and then "converted" into the hysterical symptom which in some way or other is a symbolic expression of the traumatic mental content. Honoring Freud, we find hysteria[55] in modern psychiatry under the term "Conversion Disorder," a subclass of "Somatic Symptoms and Related Disorders" (DSM-5, 2013).

Freud's theory about hysteria, like most assumptions about the workings of the unconscious, is not really verifiable. But whatever one may think about the etiology of conversion disorders, most psychologists and psychiatrists assume some kind of unconscious energy behind the symptoms. Freud had hoped to eventually develop reliable means for uncovering the unconscious mental contents causing the symptoms, for example by interpreting dreams, but as we all know (except for most psychoanalysts) he never succeeded. The unconscious is as intriguing, even mysterious today as it was in Freud's time. But, even though the weight ascribed to the role of the unconscious, both in

---

[55] It had been wrongly assumed that the symptoms, mostly found in female patients, emanate from the womb (hystera in Greek).

pathology and everyday life, differs vastly among psychiatrists and psychologists, there is no way around the general assumption that it exists and that it interacts with consciousness in the most intricate ways. Which means that all states of consciousness are never just that, states of *consciousness*. James' "stream of consciousness" is actually a stream of permanent interactions between conscious and unconscious mental processes.

Depending on the social context, there is hardly an idea, a thought, a phrase on our mind that does not carry unconscious meaning in addition to the one we experience consciously. In Freud's theorizing,[56] anxiety plays a major role in causing all kinds of unconscious mental maneuvers for keeping anxiety provoking mental contents out of the concious mind. They are generally, i.e. within as well as outside of psychoanalysis, well known as "repression" (pushing the anxiety provoking content into the unconcious), "denial" (altogether not accepting its existence), or "distortion" (changing its interpretation such that it loses ist anxiety provoking chararacter). In social psychology the most prominent theory about unconcious cognitve maneuvers affecting our thinking is the one by Leon Festinger (1957) called "theory of cognitive dissonance." A state of cognitive dissonance arises *after* one has made a decision, has taken an action, or has been exposed to information that is contrary to one's prior beliefs, feelings or values. It is experienced as a state of uneasiness one unconsciously tries to reduce ("dissonance reduction") by all kinds of cognitive maneuvers which try to make things look like there is not much of a contradiction after all. If, for example, someone decided to buy a gas guzzling SUV, which is dissonant with his engagement for environmental protection, he may tell himself (and others) that the car is a very safe one, particularly in winter when there is snow and ice, and so it may even save lives.

What do interactions between conscious and unconscious mental processes, like the ones that occur in defense mechanisms or in dissonance reduction, mean for our phenomenology of consciousness? At a closer look, it turns out that the term "interaction," meaning that there are conscious and unconscious mental processes affecting each other, is not really adequate.

---

56 Let it be noted that, without accepting all the mental mechanisms postulated by Freud, we may very well accept some of his general ideas like, for example, the defence mechanisms, all of them being mobilized unconsciously.

Think of the above environmentally conscious person with his SUV elaborating on the safety of his car. The elaborations are conscious and unconscious *at the same time*. His thoughts about safety are clearly conscious as such, but their cause, the attempt to reduce dissonance, is not. He knows what he is doing, but he does not know why. For a mental process to be called conscious both the process as such as well as its motive must be conscious. So the mental processes occurring in defence mechanisms and dissonance reduction are conceptual hybrids. There are no neatly discernible conscious and unconscious mental processes affecting each other, but both kinds of functions are interwoven in a way that creates a mental state of its own. So of the two terms James uses synonymously for describing our mental life, "stream of consciousness" and "stream of thought," the latter is the more adequate one,[57] at least if we think of "thought" as any kind of mental process, be it a conscious one, an unconscious one, or any kind of combination of the two. As so often: the closer we look at what our mental life, or psyche actually is, the more blurred, rather than more distinct the picture paradoxically becomes. So even on a descriptive level it is impossible to tell what mental processes really are, at least not with any clarity necessary for a scientific analysis. And certainly not for any attempt to emulate them.

Psychologists do not like to think much about this dissatisfactory fact; computer scientists mostly cannot even *imagine* that there might be a problem. Fundamental problems like the ones plaguing psychology are mostly unknown to physicists, engineers or mathematicians. So inevitably their ideas about human mental functions tend to be quite simplistic.

There are diverse states of consciousness, or rather consciousness-like states, caused by pathological states or induced by drugs. After brain injuries we may have all kinds of states, depending on the locus and the degree of the lesion. Conclusions about such a patient's state of consciousness are extremely difficult to draw. It turned out, for example, that an unresponsive patient, diagnosed as being in a vegetatative state long thought of as being totally unconscious, may well be completely conscious (Owen, 2017). The diagnosis "minimally conscious" (see above) leaves us completely ignorant about what is going on in the patient's mind, as do the diagnoses of neurodegenerative

---

[57] When James developed the concept, Freud had not even started to theorize about unconscious motives.

diseases like Alzheimer's. Psychiatric states, rarely diagnosed unambiguously, add many further states of consciousness to our phenomenology, like the many forms under the heading "schizophrenia." For some states like autism or some forms of depression it is even unclear whether they should be considered as pathological states in the first place. Drug-induced altered states of consciousness add a further large group of symptoms, in accordance with the many possible different chemical compositions of drugs.

The effects of the diverse pathologies, both psychiatric and neurological, and of drugs on the state of consciousness are, on the one hand, qualitatively different, but on the other, strikingly similar in some respects. Delusional states typical for schizophrenia may also be observed in patients with neurodegenerative diseases, in those with brain lesions, as well as in people under the influence of drugs and, not to forget, in normal[58] people as documented by the abundance of the most scurrilous conspiracy theories the social media are replete with.

The phenomenology of psychological states, traits and functions as presented may only give a glimpse of the daunting complexity, variety and elusiveness of the psychological world. Most of it is as little understood today as it ever was. Considering that humanity and its most brilliant minds now have been contemplating it for millennia, it may even appear inexplicable, even mystical (see below). In this perspective, the attempt to deny these mysteries by declaring man to be a machine, replicable with the help of other machines (computers), must be seen as a rather helpless one. For the sake of simplicity, it ignores the essential characteristics of mental life and in so doing does not just not add to our knowledge of it, but also keeps us from acquiring new insights into it.

---

[58] "Normal" here meaning that there is no psychiatric or neurological etiology for the condition.

Human-like computers – a scientistic delusion

As indicated at the beginning of this book, there is actually little to say about the human-like machine idea (except that it is, in Weizenbaum's terms, a mad one) if seen from the rational-scientific standpoint. From that standpoint, it would suffice to say that a computer cannot be human-like because a human is characterized by a human body and by growing up and living in a human social environment, while computers obviously lack either. Empirical evidence for the obvious can be seen in the fact that the most experienced programmers cannot make the most powerful computer conclude that the sentence "the book did not fit into the bag because it was too small" implies that the bag, not the book, was too small. The debate should simply end with that and no book should have to be written about it. But the debate as it is carried out is anything but a rational-scientifc one. It is one affected by powerful pressure groups like the AI industry with its incredible amount of money, by a large community of computer aficionados holding that computers are ultimately capable of accomplishing anything, by Hollywood, and by most of the popular scientific press.[59] They all are trying to make us believe that a new era has begun where in the not too distant future a new species starts to dominate the world, the computers. So the opposite of what Weizenbaum and other critical minds had hoped for has happened: a firm belief has been established among many, scientists included, in computers becoming human-like. But as I have tried to show there are total and unbridgeable differences between humans and computers, both structural and functional, leaving but a few particularly weak analogies between them, such as that both are able to store information, making the computer appear to have a "memory." Even if there is no such plausible analogy at hand for making the computer appear like a

---

59 Popular science does not sell if it propagates what is *not* possible. Instead the latest computer feats, irrelevant as they may be, do catch the attention.

human in other respects, like when it is purported to even have culture, there is that simple yet effective linguistic sleight of hand to have culture (or anything else) "poured into" or "fed in" the computer. At a closer look, such phrases are nonsensical, even absurd, but their use has become so common that their absurdity is simply not recognized any more. With this kind of language and its encompassing kind of thinking the impression that computers are capable of doing anything, even transform into humans, became inevitable.

In fiction, film, or art in general, where man-made humans can be created with ease,[60] such absurdities may be quite entertaining. But when they start to occupy the minds of scientists, as has actually happened, and eventually diffuse into the common mind as allegedly scientifically substantiated ideas, there is more than just a branch of science and technology gone awry. Something a lot worse has happened: our view of humanness has gone awry.

In 1748, the French philosopher Julien Offray de la Mettrie published (anonymously) *L'Homme Machine* (Man a Machine).[61] The idea is a lot older but with de la Mettrie's book it came to be seen as a characteristic one of the enlightenment, that great leap forward in mental history. But de la Mettrie's thinking shows with great clarity that the enlightenment, which rightfully bears the name, also implied, perhaps by necessity, a rather *obscuring* side effect: the belief that human reason in its strictest form, science, can eventually explain *everything* and give man the capacity to achieve anything, even to create humans. With AI's de la Mettrie-inspired project to create human-like computers,[62] we are today confronted with that obscuring side effect of rationalism, scientism, in a particularly alarming way. With irrationalism in the form of populism, antiscientific propaganda, obscurantism, belief in conspiracy theories, etc. abounding today, however, why should we be alarmed about scientism, a firm belief in the power of human reason? It is because scientism

---

[60] In film, for example, all you have to do is to have a human play a robot.
[61] More than 200 years later, the computer scientist Herbert Simon likened man to an ant, obviously thinking that ants are quite machine-like (Simon, 1969). Had he written a book about his idea it should have been titled *Man an Ant*.
[62] It appears that enthusiasm, frenzy, or hype about an idea (about the capacity of rational thought in enlightenment or about the capacity of computers in AI today) always implies a tendency to overstretch it.

is itself irrational. It is a belief-system grandiosely propagating that science can answer every kind of question, even existential ones. Scientism therefore is the patently wrong answer to irrationalism.

In 1985 one of humanity's fundamental philosophical-existential questions, "Is there free will?" was finally answered: free will does not exist (Libet, 1985). Or so it seemed. The result of Libet's experiment was confirmed some 20 years later (Son et al., 2008). Libet had found that in an experimental situation, where the subjects were told to move a finger at any time they wished to, there was a specific pattern of brain electrical activity to be observed *before* they consciously moved the finger. From that Libet somehow concluded that the movement of the finger was not caused by the subject wanting to move it. Yet critics doubted the correctness of the interpretation of the findings. And, as is not unusual in psychological experimentation, others found that free will *does* exist (Nachmias, 2015).

But whatever the latest news from this kind of science, it must be asked whether the existence of free will, relevant in many behavioral domains, can be proven or disproven by an experiment about finger movements.[63] Can a psychological experiment tell anything about free will in the first place? Considering the above-described elusiveness, even mysteriousness, of the concept of consciousness which is intricately related to free will, how can anyone proclaim a simple free will/no free will dichotomy and even dream up an experiment in order to decide upon one or the other? Such an absurd idea can only come up against the background of simplistic/mechanistic thinking, thinking that has been given up in physics long ago. It is the kind of thinking characterized by Sigmund Koch (1999) as "out of human context," a thinking that misses the essence of what psychology is all about and is thus unable to contribute anything but trivialities to our knowledge in the field.

The scientistic approach must fail with respect to all the psychological phenomena that cannot be dealt with on a purely scientific basis. Ethical behavior for instance requires free will and it should not surprise us that the attempts to scientifically prove or disprove its existence are hopelessly ambiguous. Moral principles like the ones laid down in the American Declaration

---

[63] Typically the free-will experimenters took physics as their scientific model, where the existence or nonexistence of, for example, a certain boson or ether *can* be decided upon experimentally.

of Independence, for example, are not based on any scientific evidence. They are, as the authors of the document put it, "self-evident." They used the phrase in order to convey that there are no specific reasons, and certainly not any scientific ones, to be brought forward in support of the "truths" proclaimed, like the one "that all men are created equal."

To eventually leave ethical decisions to machines would be the most disastrous consequence of the belief in human-like machines. In medicine, decisions about different treatments (or no treatment), which clearly involve ethical aspects, must not seldom be made. Among all the behavioral domains where ethical decisions must be made, autonomous weapons, if left with ethical decisions about life and death of combattants and civilians, would perhaps be the crassest example of the disastrous consequences of leaving ethical decisions up to machines.[64] Under any circumstances it must be clear that machines cannot possibly make ethical decisions the way humans can, i.e. decisions that consider the human context.

How rational and scientific thought, when applied to ethics, fails to grasp the essence of ethical behavior has inadvertently been shown by the great 18th century philosopher and moral theorist Immanuel Kant, who at the same time must be considered the father of modern concepts of race and scientific racism (Sussman, 2014, p. 27). Racism was common among philosophers and scientists at the time and still was a century later (Darwin thought blacks to be on a lower stage of evolutionary development as to their intellectual and moral faculties).[65] But the conclusions drawn by Kant and Darwin from their view of blacks could not be more different. Darwin was a committed abolitionist who saw slavery as a "great crime" and was prepared to take serious risks in voicing his opinion, for example when, on the ship on which he travelled as a naturalist, he confronted the captain of the ship, a staunch defender of slavery. Kant, the great teacher of moral and ethics, came to a different

---

[64] I held my breath for a moment when I read that in the 50s of the last century a hype about mathematical models of decision making prompted scientists to propose to leave decisions about the use of nuclear weapons to computers. Humans were deemed too unreliable and too irrational for such things. The argument is often heard today in favor of self-driving cars.

[65] It still exists a further century later (e.g. Rushton, 1995) even though not being common any more.

conclusion. According to his "theory" about the races (Europeans, Asians, Africans, and Native Americans), the intellectual and moral incompetence of blacks meant that they could only be educated as servants (slaves) and had to be kept under control by severe punishment. He even explained how to properly beat them with split bamboo canes (p. 28). So the great thinker of the enlightenment, protagonist of rationality and reason, found that the utmost instance of inhumanity – putting humans into slavery and beating them severely – was a morally defensible, even necessary act.[66] So much for the rational, scientific approach to moral behavior.

Ethical value systems, like the one underlying the above famous document, are usually based on the belief in some supernatural entity or power often called "god." So the authors of the Declaration of Independence in their concluding sentence speak of a "firm reliance on the Protection of Divine Providence." Today, however, many people, certainly more than at the end of the 18th century, no longer believe in anything supernatural, be it something personal like the god of the Christians or something more diffuse like a spirit pervading the universe. And to the loss of belief in anything spiritual or supernatural they respond in the same way as did de la Mettrie, i.e. by resorting to a simple materialism/determinism.[67] With respect to the human mind, this attitude, termed "scientism" today (e.g. Comfort, 2012), implies the belief that all mental processes are based on material (mostly neural) functions, which is correct, and that they therefore can ultimately be *explained* by the proper study of these functions, which is wrong. Proponents of this scientistic attitude see it as the only alternative to spiritualism, the belief in things supernatural.

As we have seen, this attitude remains undeterred by all the failures to scientifically explain mental states and functions beyond just pointing to material *correlates* of those functions, which does not explain anything. As

---

[66] I think that Kant should be reevaluated with respect to his racist and inhuman attitude. As Matthew Hachee puts it: "[…] the image of Kant […] is one excessively sanitized […] as a result of a tradition conveniently blind to its own racism" (p. 30).

[67] Some progress in this matter, even though no epistemological one, may be seen in that a book with the content of de la Mettrie's *Histoire naturelle de l'âme* (A natural history of the soul) would not be condemned and burned by court order today as was de la Mettrie's book.

we have seen, the state of affairs with respect to the attempt to scientifically explain the all-important concept of consciousness, for example, must simply be termed chaotic. So what should a rational alternative to spiritualism, one that does not contradict scientific method and evidence, be? In contrast to scientism the alternative would be based on the insight that there are questions, even quite rational ones, that cannot be answered by science, thus always leaving realms of the unknowable and even the mysterious. Note that the term "mysterious" does not necessarily imply supernatural entities. As an example of mysticism without any supernatural entities take the attitude of Joseph Conrad whose description of tropical nature typically conveys a mystical atmosphere. Not postulating anything supernatural he wrote: "The world of the living contains enough marvels and mysteries as it is; marvels and mysteries acting upon our emotions and intelligence so inexplicable that it would almost justify the conception of life as an enchanted state" (Hitchens, 2007, p. 123). It may even happen, as it did in physics (particle physics and cosmology), the science perhaps most closely tied to objective, empirical evidence and logical inference, that ordinary common-sense notions (of space, time, matter, and energy) are no longer valid. New insights always create new questions. As to the cosmos, for example, my impression, when following the latest popular scientific publications, is that both its origin and course must be thought anew every some months (e.g. Giddings, 2019).[68]

Concerning the human mind, consensus seems to exist that it must have a physical basis, a physical substratum. Yet there is no consensus about what follows from this as to how to study mental functions. And it seems that most of those, who have concluded (understandably) that the study of those functions must pursue the study of the underlying physical, mostly neural, functions, have not given all too much thought to the implications of that project.

Critics of the project have pointed to the complexity of the substratum, essentially (not exclusively) the brain with its about 100 billion neurons, each of them having several thousand connections (synapses) to other neurons with different chemical processes happening at different synapses. It is to be added that the brain interacts with highly complex neural structures outside the brain and ultimately with the whole body. But, speaking of the substrate

---

[68] The article postulates: "Our very notion of space and time, which underlie the rest of science, appears to require significant revision" (p. 48).

of psychological functions, there is a further aspect, actually a psychological platitude, which makes the study of mental functions via neural ones look patently absurd: mental functions differ qualitatively between individuals. As we have seen, even such relatively simple mental functions as the memorizing of numbers may be performed in totally different ways. In the context of complex cells, we have seen that different ones may develop depending on the mental requirements different people are confronted with.[69] The emotional response to an identical social situation may be totally different depending on the emotional history of the individual. There are, of course, similarities between people as to mental functions, but the more complex these functions (think of a situation where a moral decision must be made), the more idiosyncratic (qualitatively different for different people) they are. With qualitative differences existing between mental functions there must be concomitant qualitative differences between the corresponding neural substrates, which precludes any general neural explanation of those mental functions.

As mentioned, it is misleading to present complex cells for visual features common to all people (or animals) in psychological textbooks because it gives the wrong impression that there must be complex cells for all kinds of *psychological* functions too. This false impression may have misled many a neuropsychologist into believing in the flawed project of studying psychological functions via neural ones.

As to the physical basis of psychological functions, the typical question "How does the brain work?" is therefore one of the many unanswerable ones.[70] The neural basis of mental functions, even though it obviously exists, will remain a mystery on a-priori grounds. The scientistic belief that the question will eventually be answered, naïve as it is, has cost and will cost billions, both dollars and hours invested by intelligent people who could do something more reasonable instead.

So the alternative to the belief in supernatural entities should not be scientism, i.e. a simple deterministic attitude, but the insight that for many human existential question mysticism, the feeling of unexplainability, will

---

[69] "The yellow volkswagen detector" has jokingly been postulated decades ago for people who, for whatever reasons, have ample experience with such cars.

[70] The answer, "As it pleases the individual brain," even though not really illuminating, is the only one we can give.

stay with us. Admittedly this insight will make people trained in rational explanation and demystification feel uneasy. Rational people have felt that uneasiness ever since Socrates' insight into the limits of our capacity to explain the world some 2500 years ago. But it is a deeply *rational* insight.

Not everyone with a mechanistic-deterministic view of human nature must end like de la Mettrie, the philosopher of *L'Homme Machine*. The biological mechanisms of that machine being regulated to just seek pleasure and flee pain, he, who lived accordingly, probably died from overeating at the age of just 42 years. But there will probably be subtler effects of today's man-a-machine attitude, not necessarily ones in the form of people's conscious conclusions about how to live. As to the workings of democracy, for example, the mechanistic-deterministic attitude is an extremely naïve one. You may install democracy with its corresponding institutions, but that does not lead automatically, i.e. by social-political mechanisms to the intended goal, a free and just community. If that community is not to degenerate into corruption, concentration of power, and the dictatorship of a few, it takes the democratic *spirit* according to John F. Kennedy's formula "Don't ask what your country can do for you, ask yourself what you can do for your country". There are, of course, diverse factors (political, sociological, psychological, etc.) affecting the functioning of democracy, which can be studied scientifically and eventually be modified in order to stabilize or advance democracy. But the spirit of democracy eludes and will elude scientific scrutiny. It will at times light up, as, for example, with Kennedy's words, and make people somehow *feel* what it is. But all we know rationally about it is that it is a conglomerate of ideas like freedom, justice, equality, or humanism which have evolved over millennia with crucial historical steps like the constitutions of the ancient Greek city states, the Roman Republic (not a democracy), or the American and the French revolutions. Concrete as the effects of that spirit may be (in the concrete rules for voting, for example), it will always stay a myth in the sense that we cannot scientifically define and explain it. A myth not necessitating the belief in any supernatural entities, however.

Scientific progress, for example in genetics, stemcell research, reproduction medicine, or military technology produces fascinating but often monstrous possibilities which have to be decided upon, not on the basis of scientifically validated strict rules, but of a general humanistic attitude. Science just supplies the facts on the basis of which those decisions are made. I would not dare to define what a humanistic attitude is, as little as I would try to define art, or consciousness for that matter. All I know is that it exists as long as people believe in it and that it is, and always has been, in peril. In the 20th century it was jeopardized by the powerful pseudo-scientific ideologies of Fascism, with its belief in biological determinism, and Communism, with its belief in social determinism. Today and in the foreseeable future, scientism may turn out to present the most powerful peril, its ideology mistaking the essence of democracy, its spirit, for something irrational that must be replaced by science.

In the last chapter of *Computer Power and Human Reason* titled "Against the Imperialism of Instrumental Reason" Weizenbaum draws practical and philosophical conclusions from his foregoing critique of the mental state of the community of computer scientists, a critique quite reminiscent of the one Sigmund Koch articulated about the mental state of psychologists who see themselves as natural scientists, characterizing their research as "out of human context." He writes: "The various forms of human and social engineering we have discussed here do just that (dehumanize) in that they circumvent all human contexts" (p. 266).

The gist of both Weizenbaum's and Koch's critique is the loss or absence of humanism in sciences dealing with the human psyche, i.e. in psychology proper (as criticized by Koch) and in any other science addressing psychological functions, like in a computer science which tries to emulate such functions (as criticized by Weizenbaum). Humanism being a non-scientific concept or set of ideas, there can be no *scientific* definition of it. But the definition given, for example, in *Webster's Collegiate Dictionary* (2009), incomplete as it may by necessity be, still catches the essence: "A philosophy that usu. rejects supermaterialism and stresses an individual's dignity and worth and capacity for self-realization through reason." As the American Declaration of Independence states about the human rights it proclaims, humanism as the philosophy for a fair, just and free society is "self-evident."

Two brilliant minds, both highly respected in their respective fields, criticizing their field in a fundamental way, creates a rather puzzling, even absurd situation. Not liking absurdities, most scientists have avoided discussing the critique, many of them because that critique was "just" a philosophical and, as they saw it, a negligible one. The effect of the critique has been limited, if not absent, in both fields. In psychology there actually has even been a presidential initiative by the American Psychological Association to stylize the field *more* in the direction of the STEM (science, technology, engineering, mathematics) fields (Weir, 2014, 2015), which mostly proceed successfully without having to give much attention to any human context. That initiative must make us doubt that the call for a human context has been heard among psychological scientists, at least among those who represent them, like the American Psychological Association. In the same vein there has been little opposition from the psychological scientific community against the most scientistic project of all, the creation of human machines. As to be expected, there was none from the AI community. The opposite of what Koch and Weizenbaum had intended has happened in their respective fields. Why?

The answer is as simple as it is disillusioning. Both fields have been successful by proceeding *out of* human context, success, to be sure, here not meant in terms of scientific progress but of prestige gained. Psychology has presented itself successfully as one among the natural sciences, allegedly offering a solid pool of fundamentals, successfully applicable to problems quite like those an engineer learns about in his studies. And although AI has proven to be in no need to gain prestige from projects trying to build human-like machines, as it has shown it can effectively be used to solve all kinds of problems[71], it unfortunately has not resisted the temptation to gain attention (and further prestige) from the claim, seen as particularly fascinating by most people, that it is able to create human psychological functions in computers and ultimately able to create humans and even superhumans, whatever the latter may mean. And in so doing, it has necessarily proclaimed the scientific idea of omnipotence.

So what can be done in a situation which must be a frustrating one for all scientists with a realistic view of the human psyche and without the belief in the omnipotence of computers? Realistically there is one thing we may predict with near certainty. The believers in human-like computers will not some day be befallen by the insight that their belief is an absurd one. They have avoided that insight successfully for some 60 years. Empirical evidence from more than six decades and a short glance at the interest structure of those

---

[71] If AI, for example, produces a robot which helps handicapped people move around in their homes, this robot will typically be made to look somewhat human-like, but it is quite unnecessary to make people believe that it has human-like thoughts or feelings.

involved in the research tells us that grandiose predictions about future feats will be forthcoming on a weekly basis. The next AI winter will probably come, but it will be overcome as long as that feeling of omnipotence and the scientistic attitude will endure. When the deep-learning bubble will have turned out to be exactly that, a bubble, the next one will already have been created, most probably the belief that quantum computers will solve all problems, long before it will emerge whether quantum computers can be of any use at all in the first place. The mere mentioning of the claim that quantum computers will be a billion (trillion, quadrillion, …) times faster than the most powerful computers today, combined with the firm belief that scientific progress is mainly a matter of computer power, will probably convince people, scientists included, that *they* will do the trick. As psychology has shown, overselling its achievements can be made the raison d'être of a whole field of science for more than a century.[72]

Nevertheless, I hope that scientific rationality will eventually prevail, meaning that AI will feel committed to strict scientific standards where wild and untestable speculations are not permitted and absurdities like culture being poured into a computer are seen as just that.

But this can only happen on a broader mental-historical background, with the abandonment of, or at least a vivid skepticism about, scientism with its belief in the computer assisted omnipotence of science, including the possibility of humans creating humans. While Frankenstein's monster and its computer-controlled descendants may, and for reasons of entertainment should definitely be allowed to, further populate the world of imagination, we need not, as I have tried to show in this book, be afraid that they might populate the real world.

But the thought that, in the course of a further mental-historical shift toward scientism, those populating the *real* world, the ones we call humans, might be seen as nothing but machines, programmed by modern Doctor Frankensteins, should frighten us as much as it frightened those who ordered to burn de la Metrie's book.

---

[72] Intelligence tests, the first one constructed in 1904, for example, are seen in a wide public as scientifically based tools, even though they have demonstrably not lived up to the claim that they can make any substantive predictions about success in any mental performance domain (e.g. Murdoch, 2007).

Acknowledgements

I wish to thank Alfred Weber for many linguistic suggestions, and Frauke Ewert for her moral support in the face of a huge majority of colleagues firmly believing in human-like computers.

Literature

Adee, S. (2016). Will AI's bubble pop? *New Scientist, 231(3082)*,16–17.
Alexander, F. (1950). Psychosomatic Medicine. New York: W. W. Norton & Co. Inc.
Banino, A. et al. (2018). Vector-based navigation using grid-like representations in artificial agents. *Nature, 557,* 429–433.
Blackmore, S. (2018). Decoding the Puzzle of Human Consciousness. *Scientific American, 319(3),* 41–45.
Blackmore, S., & Troscianco, E. T. (2018). Consciousness: An Introduction (3rd ed.). London: Routledge.
Brooks, M. (2020). Here. There. Everywhere? *New Scientist,* 2 May 2020, 40–44.
Carr, N. (2010). The Shallows. London: Atlantic Books.
Collins, H. (2018). Artifictional Intelligence. Cambridge, UK: Polity Press.
Comfort, N. (2012). The Science of Human Perfection. New Haven: Yale University Press.
Crick, F. (1994). The Astonishing Hypothesis. London: Simon & Schuster.
Daly, M. & Wilson, M. I. (1982). Homicide and Kinship. *American Anthropologist, 84,* 372–378.
Davis, E. (2016). How to Write Science Questions that are Easy for People and Hard for Computers. *AI Magazine, 31(1),* 13–22.
Dennet, D. C. (2017). From Bacteria to Bach and Back: The Evolution of Minds. W. W. Norton.
Domingos, P. (2018). Our Digital Doubles. *Scientific American, 319(3),* 80–85.
DSM-5 (2013). Diagnostic and Statistical Manual of Mental Disorders (5th ed.). Washington, DC: American Psychiatric Publishing.
Editorial (2017). Intelligence test. *Nature, 545(7655),* 385–386.
Efron, R. (1990). The Decline and Fall of Hemispheric Specialization. Hillsdale, NJ: Lawrence Erlbaum.
Festinger, L. (1957). A Theory of Cognitive Dissonance. Stanford, CA: Stanford University Press.
Geddes, L. (2019). Height's 'missing heritability' found. *Nature, 568, (7753),* 25 April 2019, 444–445.
Giddings, S. B. (2019). Escape from a black hole. *Scientific American, 321(6),* 42–49.
Goldstein, E. B. (2014). Sensation and Perception. Belmont, CA: Wadsworth.
Harpaz, Y. (1999). Replicability of Cognitive Imaging of the Cerebral Cortex by PET and fMRI: a survey of recent literature. http://www.yehouda.com/replicability.html.

Graziano, M. (2019). What is consciousness? *New Scientist,* 21 September 2019, 34–37.
Heaven, D. (2018). AT's lost genius. *New Scientist, 238(3180),* 40–41.
Heaven, D. (2019). Deep trouble for deep-learning. *Nature, 574(7777),* 163–166.
Heider, F. (1958). The psychology of interpersonal relationships. New York: Wiley.
Hickok, G. (2014). The Myth of Mirror Neurons. New York: W. W. Norton & Company.
Hitchens, C. (2007). The Portable Atheist. Philadelphia, PA: Da Capo Press.
Hubel, D. H. & Wiesel, T. N. (1962). Receptive fields, binocular interaction and functional architecture in the cat's visual cortex. *Journal of Physiology, 160,* 106–154.
Hull, C. (1943). Principles of Behavior. New York: Appleton Century Crofts.
Hutson, M. (2021). The language machines. *Nature, 591(7848),* 22–25. (4 March 2021).
Koch, C. (2018). What is Consciousness? *Nature, 557, (7704),* S9-S12.
Koch, C. (2018). What is Consciousness? *Scientific American, 318(6),* 57–60.
Koch, S. (1999). Psychology in Human Context. Chicago: The University of Chicago Press.
Koch, S. (1999). Psychology's Bridgman versus Bridgman's Bridgman: A Study in Cognitive Pathology. In D. Finkelman & F. Kessel (Eds.). Sigmund Koch – Psychology in Human Context. Chicago: The University of Chicago Press.
Köhler, W. (1927). The Mentality of Apes. New York: Humanities Press. (Original: Intelligenzprüfung an Menschenaffen, 1917).
Konorski, J. (1967). Integrative Activity of the Brain. Chicago: University of Chicago Press.
Kroeber, A. L. (1948). Anthropology. Harcourt.
Kurzweil, R. (2005). Human 2.0. *New Scientist, 187(2518),* 32–37.
Kurzweil, R. (2012). How to Create a Mind: The Secret of Human Thought Revealed. New York: Viking Penguin.
Levesque, H. (2017). Common sense, the Turing Test, and the Quest for Real AI. Cambridge, MA: MIT Press.
Libet, B. (1985). Unconscious cerebral initiative and the role of conscious will in voluntary action. *The Behavioral and Brain Sciences, 8,* 529–539.
Loftus, E. F. (1993). The reality of repressed memories. *American Psychologist, 48,* 518–537.
Luria, A. (1973). The Working Brain. Harmondsworth: Penguin.
Marcus, G. (2017). Am I human? *Scientific American, 316(3), 52–55.*
Marcus, G., & Davis, E. (2019). Rebooting AI. New York: Pantheon Books.
Melzack, R., & Wall, P. D. (1965). Pain mechanisms: a new theory. *Science, 150,* 971–979.
Mettrie, de la, J. O. (1748, 1921). L'Homme Machine. Paris: Solovine.
Murdoch, S. (2007). IQ – A Smart History of a Failed Idea. Hoboken, NJ: John Wiley & Sons Inc.
Nachmias, E. (2015). Why we have free will. *Scientific American, 312(1),* 65–67.
Neisser, U. et al. (1996). Intelligence: Knowns and unknowns. *American Psychologist, 51,* 77–101.
Nisbett, R. E., Aronson, J., Blair, C., Dickens, W., Flynn, J., Halpern, D. F., & Turkheimer, E. (2012). Intelligence – New Findings and Developments. *American Psychologist, 67(2),* 130–159.
Owen, A. (2017). Into the Gray Zone. New York: Scribner.

Penfield, W., & Perot, P. (1963). The brain's record of auditory and visual experience. *Brain, 86,* 595–696.
Penrose, R. (1989). The Emperor's New Clothes. New York: Oxford University Press.
Plomin, R. (2018). Blueprint. Allen Lane.
Reed, C. (2021). Argument technology for debating with humans. *Nature, 591(7850),* 373–374 (18 March 2021).
Rizzolati, G. et al. (1996). Localization of grasp representation in humans by PET: I. Observation versus execution. *Experimental Brain Researh, 111,* 146–152.
Rushton, J. P. (1995). Race, Evolution and Behavior: A Life History Perspective. New Bruswick, NJ: Transaction.
Russell, B. (1959). My Philosophical Development. London: George Allen & Unwin Ltd. Chapter 12.
Savelli, F. & Knierim, J. (2018). AI mimics brain codes for navigation. *Nature, 557,* 313–314.
Shen, H. (2013). Brain storm. *Nature, 503(7474),* 26–28.
Siemons, M. (2019). Wir Cyborgs. *Frankfurter Allgemeine Wochenzeitung, 31,* 33.
Simon, H. A. (1969). The Sciences of the Artificial. Cambridge, Mass.: The MIT Press.
Skinner, B. F. (1957). Verbal Behavior. New York: Appleton Century Crofts.
Snaprud, P. (2018). The Consciousness Wager. *New Scientist,* 23 June 2018, 28–31.
Snowden Jr., F. M. (1983). Before Color Prejudice. Cambridge, MA: Harvard University Press.
Sohn, E. (2019). Decoding consciousness. *Nature, 571 (7766),* S2-S5.
Soon, C. S., Brass, M., Heinze, H.-J., & Haynes, J.-D. (2008). Unconscious determinants of free decisions in the human brain. *Nature Neuroscience, 11,* 543–545.
Sukel, K. (2012). The amazing memory marvels. *New Scientist, 215(2878),*34–37.
Sussman, R. W. (2014). The Myth of Race. Cambridge, MA: Harvard Univerity Press.
Tucker, W. H. (2002). The Funding of Scientific Racism. Urbana and Chicago: University of Illinois Press.
Turk, D. C. (1990). Customizing treatment for chronic pain patients: Who, what, and why. *Clinical Journal of Pain, 6,* 255–270.
Uttal, W. R. (2001). The new Phrenology. Cambridge, Mass.: The MIT Press.
Velden, M. (2010). Biologism – the Consequence of an Illusion. Göttingen: V&Runipress.
Velden, M. (2014). Brain Death of an Idea – the Heritability of Intelligence. Göttingen: V&Runipress.
Velden, M. (2016). Psychology – A Study of a Masquerade. Göttingen: V&Racademic.
Velden, M. & Vossel, G. (1985). How can skin conductance responses increase over trials while the corresponding resistance responses decrease? *Physiological Psychology, 13,* 291–295.
Weir, K. (2014). Translating psychological science. *Monitor on Psychology, 45(9),* 33–36.
Weir, K. (2015). Truth in advertising. *Monitor on Psychology, 46(3),* 36–38.
Weizenbaum, J. (1976). Computer Power and Human Reason. London: W. H. Freeman and Company.

Wertheimer, M. (1922). The general theoretical situation. In W. D. Ellis (Ed.). A sourcebook of Gestalt Psychology. New York: Harcourt Brace, 1938, 12–16. Also in: Watson, R. I. (1979). Basic Writings In the History of Psychology. New York: Oxford University Press, 291.

Zeki, S. (1993). A Vision of the Brain. Oxford: Blackwell Scientific Publications.

Index

agnosias
- definition, etiology of 46, 47, 82
- group- and person-specificity of 47

AI winter
- first 31
- next 32, 58

analogies
- definition of 29
- misleading use of in AI research 29, 30
- as raison d'être of human-like-computer idea 29, 30

anthropomorphization
- and belief in human-like machines 32, 33
- and panpsychism 97

attribution theory
- and the unconscious 82

autism
- and mirror neurons 20, 21

behaviorism
- as abandoned approach in psychology 16–18
- renaissance of in AI research 53, 54, 69

brain
- as a computer 33, 55–57
- relation to body and environment 71, 72
- as inadequate substitute for the human 71, 72

cerebral cortex
- and complex cells 44
- and the agnosias 46, 47
- and psychological processes 50

cognitions
- interactions with emotions 86–88
- role in theories of emotion 86
- role in psychotherapy 86

Crick, F.
- and consciousness 92

complex cells
- misleading presentation of 45
- person- and group specificity of 46, 47

computer metaphor (the brain as a computer)
- role in human-like-computers idea 55, 57
- inventor of 55

consciousness
- theories of 96–98
- role in jurisdiction 93
- phenomenology of 98–105
- role of the unconscious in 99, 102, 103
- states of 100, 101

constructs (hypothetical)
- and reification/localization 21
- as subject matter of psychology 21

Darwin, C.
- and slavery 110

Darwinism
- as ideology 77

deep-learning
- definition of 48
- and human learning 48, 49
- and next AI winter 58

determinism
- and free will 109

– and modern physics 112
– and spiritualism 112
dissonance theory
– and the unconscious 103

Efron, R.
– and hemispheric specialization program 23
ELIZA
– as Turing Test 65
– and anthropomorphization 32, 65
– and language 65
– and psychotherapy 32
emotions
– biological correlates of 85, 86
– and cognitions 86, 87
– and pain 86, 87
enlightenment
– obscuring side effect of 108
– and scientism 108

Fechner, G. T.
– psychophysical law of 40, 41
– and elementism 39
Freud, S.
– and the unconscious 102
– role in psychosomatics of 19
– and defence mechanisms 103
– and hysteria (conversion symptoms) 102

Galton, F.
– and the heritability of mental functions 14
Gestalt Psychology
– as opposition to behaviorism 17
– and insight 49
– and experiments in learning 49, 81
– and deep-learning 49

HAL (from "2001 – A Space Odyssey")
– as grandiose AI prediction 31
Hawking, S.
– and computer metaphor of the brain 55

Hickok, G.
– and the debunking of the mirror neuron hype 20
– and hemispheric specialization 23
Hollywood
– as powerful propagator of human-like machines 33
humanism
– definition of 114, 115
– as spiritual basis of democracy 115
Hull, C. L.
– flawed psychological program of 17
– and use of mathematics in psychology 17
– and delusional comparison to Copernicus, Galilei, and Newton 17

intelligence
– heritability of 14, 15
– in humans vs. in machines 27
– and chess 28
– and intelligence tests 117

James, W.
– and stream of consciousness (stream of thought) 38, 73
– and Fechner's law 40, 41
– and elementism 39

Kant, I.
– and slavery 110, 111
– and scientific racism 110, 111
Koch, S.
– critique of academic psychology by 115
– and psychology in human context 109
– and scientism 109, 115
Köhler, W.
– experiments with apes 49
– and insight (Aha experience) 49
Kurzweil, R.
– garbled language of 56
– outlandish predictions of 55
– and computer metaphor 55
– and language comprehension 56

language
- Kurzweil's view of 56
- Skinner's teory of 17
- in computers 56, 57
lateralization
- as modern phrenology 22, 23
learning
- and conditioning 90, 91
- Gestalt view of 49
- Human learning vs. deep-learning 49, 92
localization (of mental functions)
- as modern phrenology 23
- replication attempts of studies about 22
- methodological problems of 22
- and reification 21

memes
- analogy to genes 97
- in theory of consciousness 97
memory
- factors affecting 78, 79
- for numbers 74
- for concepts 75–77
Mettrie, Offray de la
- and scientism 111, 114
- as precursor of human-like-computers idea 10, 108
- philosophy of 114
Minsky, M.
- outlandish predictions of 31
mirror neurons
- hype about 20
- and autism 20, 21
- debacle of field of research 20
mysticism
- and supernatural entities 113
- as response to scientifically unsolvable problems 112, 113
- as a rational attitude 112, 114

neuron
- as compared to computer elements 89, 90

operationalism
- use in neurology 99, 100

pain
- as an emotion 86
- and cognition 86
perception
- role of complex cells in 44–47
- and percepts 50
- and visual-gnostic categories 82, 83
- and visual agnosias 82
physics
- as model science for psychology 39
psychosomatics
- role of Freud in 19
- dogma of organic causation in 19

racism (in science)
- and heritability research 16
- and Immanuel Kant 110, 111
- absence of in Roman antiquity 83

scientism
- definition of 108
- as an ideology 107–117
- and free will 109
self driving cars
- as grandiose announcement 31
- non-existence of 31
Skinner, B. F.
- and language 17
- and Skinner-Chomsky controversy 17
socialization of a machine
- as a mad idea 7, 9
stream of consciousness (stream of thought)
- definition of 38
- as concept largely ignored in modern psychology 38
- as primary object of psychological science 38
surrealism (dadaism)
- similarity to AI speak 10, 67

thinking
- as encompassing term for mental functions 91, 92, 104
- and Descarte's "cogito ergo sum" 91
- and James' stream of thought 91

Turing Test
- definition of 65
- creation of 66
- scientific value of 68, 69

unconscious, the
- and defense mechanisms 103
- relation to consciousness 103, 104

- and attribution theory 82
- and dissonance theory 103
- an dissociation 78

Uttal, W. R.
- and localization of mental functions 22

Watson, J. B.
- and behaviorism 16, 17

Weizenbaum, J.
- and ELIZA program 32, 65
- "madness of our time" statement by 7, 9, 24, 25
- and scientism 115
- and Sigmund Koch 115, 116

Das Signet des Schwabe Verlags
ist die Druckermarke der 1488 in
Basel gegründeten Offizin Petri,
des Ursprungs des heutigen Verlags-
hauses. Das Signet verweist auf
die Anfänge des Buchdrucks und
stammt aus dem Umkreis von
Hans Holbein. Es illustriert die
Bibelstelle Jeremia 23,29:
«Ist mein Wort nicht wie Feuer,
spricht der Herr, und wie ein
Hammer, der Felsen zerschmeisst?»